T0225180

PSpice for Circuit Theory and Electronic Devices

© Springer Nature Switzerland AG 2022

Reprint of original edition © Morgan & Claypool 2007

PSpice for Circuit Theory and Electronic Devices
Paul Tobin

ISBN: 978-3-031-79754-5   paperback
ISBN: 978-3-031-79755-2   ebook

DOI: 10.1007/978-3-031-79755-2

A Publication in the Springer series
SYNTHESIS LECTURES ON DIGITAL CIRCUITS AND SYSTEMS #7

Lecture #7
Series Editor: Mitchell A. Thornton, Southern Methodist University

Library of Congress Cataloging-in-Publication Data

Series ISSN: 1932-3166   print
Series ISSN: 1932-3174   electronic

First Edition
10 9 8 7 6 5 4 3 2 1

PSpice for Circuit Theory and Electronic Devices

Paul Tobin
School of Electronic and Communications Engineering
Dublin Institute of Technology
Ireland

SYNTHESIS LECTURES ON DIGITAL CIRCUITS AND SYSTEMS #7

ABSTRACT

PSpice for Circuit Theory and Electronic Devices is one of a series of five PSpice books and introduces the latest Cadence Orcad PSpice version 10.5 by simulating a range of DC and AC exercises. It is aimed primarily at those wishing to get up to speed with this version but will be of use to high school students, undergraduate students, and of course, lecturers. Circuit theorems are applied to a range of circuits and the calculations by hand after analysis are then compared to the simulated results. The Laplace transform and the s-plane are used to analyze CR and LR circuits where transient signals are involved. Here, the Probe output graphs demonstrate what a great learning tool PSpice is by providing the reader with a visual verification of any theoretical calculations. Series and parallel-tuned resonant circuits are investigated where the difficult concepts of dynamic impedance and selectivity are best understood by sweeping different circuit parameters through a range of values.

Obtaining semiconductor device characteristics as a laboratory exercise has fallen out of favour of late, but nevertheless, is still a useful exercise for understanding or modelling semiconductor devices. Inverting and non-inverting operational amplifiers characteristics such as gain-bandwidth are investigated and we will see the dependency of bandwidth on the gain using the performance analysis facility. Power amplifiers are examined where PSpice/Probe demonstrates very nicely the problems of cross-over distortion and other problems associated with power transistors. We examine power supplies and the problems of regulation, ground bounce, and power factor correction. Lastly, we look at MOSFET device characteristics and show how these devices are used to form basic CMOS logic gates such as NAND and NOR gates.

KEYWORDS

Cadence Orcad PSpice V10.5, Ohm's law, Kirchhoff's laws, Thévenin and Norton theorems, Mesh and nodal analysis, Laplace, transients, transfer functions, resonance, transformers, power supplies, ground bounce, operational amplifiers, power amplifiers.

I would like to dedicate this book to my wife and friend, Marie and sons Lee, Roy, Scott and Keith and my parents (Eddie and Roseanne), sisters, Sylvia, Madeleine, Jean, and brother, Ted.

Contents

Preface

Many years ago, I discovered how electronic simulation helped students come to grips with difficult engineering concepts. Earlier simulation software used cumbersome circuit netlists but nevertheless showed me how it helped students gain an intuitive circuit design sense. PSpice evolved along with the Windows environment to produce, in my opinion, a very powerful teaching and learning tool for accessing a whole range of difficult areas such as circuit theory, electronics, telecommunications and digital signal processing (DSP). This book, and my other fours books, grew from laboratory exercises and projects given to my student over the last twenty years.

An unfortunate trend in engineering education throughout the world has been to reduce analogue circuit design and circuit theory when considering new course syllabi. This is due, in part, to the ever-growing software-based technology such as the Open Systems Interconnection (OSI) networking model and associated protocols, C, C++, Java etc. Something has to go and unfortunately it seems to be some important basic principles. Students find digital circuits and DSP much easier to understand than analogue circuits and hence students tend to 'cherry pick' the easier topics ending up with a poorer overall understanding of engineering design. This is leaving the engineering recruitment market suffering from a lack of analogue design engineers. Good analogue circuit design is a combination of circuit analysis, an intuitive feel for electronic design and engineering problem solving obtained from experience. PSpice comes to the rescue with all these problems and helps students develop an intuitive design sense in a much shorter time.

This book is a combination of textbook and laboratory manual and contains worked examples with sufficient theory to enable the reader to compare simulation results to hand calculations. Exercises at the end of each chapter are partly worked to encourage the student to finish to completion. Lecturers should find the book as a valuable source for examination questions (loud groan from all), laboratory work, student projects and lecture material. It should also be very useful to second-level high school teachers where electronic technology has been introduced into the curriculum for some years. The book contains eight chapters covering topics from DC, AC and electronic devices. Chapter 1 introduces PSpice version 10.5 using a very simple DC circuit. Chapter 2 examines fundamental electric circuit principles and circuit theorems applied to DC and AC networks. In chapter 3, we look at the Laplace transform applied to first–order CR and LR switching circuits where the simulation outputs of currents and voltage at different times may be compared to hand calculations. Chapter 4 continues with more s-plane circuits and examines Butterworth and Chebychev transfer function. Chapters 5 and 6 analyses and simulates, AC circuits and applies circuit theorems such as Thévenin's theorem, mesh and nodal analysis to a range of circuits, including series and parallel resonant circuits. In Chapters 7 we plot electronic device characteristics in order to design circuits using measured device parameters from the characteristics. In the last chapter we examine operational and power amplifiers and a brief visit to CMOS devices and logic gates.

ACKNOWLEDGMENTS

I was introduced to circuit theory and electronics when I attended, many years ago, a very comprehensive series of lectures on these topics given by a fine lecturer and retired head of our department, Chris Cowley, so my thanks to him now many years later. I should also thank my students, past and present for inadvertently proof reading my books.

CHAPTER 1

Introduction to PSpice and Ohm's Law

1.1 LAYING OUT A SCHEMATIC

The latest version of Orcad® PSpice® 10.5 is started by selecting **Capture CIS** from the **Windows Start/Programs** menu. This version has many different features from versions 8 and 9 and those of you who operate still with these versions will find it a little difficult at the start but persevere its worth it. A new project file must be created before a schematic is created. The project file is not a design file but project management and is initiated by selecting the small folded white sheet icon at the top left-hand corner of the display shown in Fig. 1.1.

Fill in the **Name,** select **Analog or Mixed A/D** and specify a **Location** for the file. Press **OK** and a further sub-menu opens requesting you to create a blank project as shown in Fig. 1.2. (If you are using the demo version, then several messages will pop up to tell you about the limitations.)

This produces an empty schematic area called **Page 1** shown in Fig. 1.3 where component symbols, selected from the library, are placed. However, if you select **Create based on** etc., then all previously used libraries with that project will be loaded. There are several operating techniques for creating and simulating a schematic. One may use the top icon toolbar with drop-down menus **File**, **Edit**, **Draw** etc. and the icon toolbar on the right-hand side of the page area. Windows short-cut keystrokes **Ctrl C**, **Ctrl V**, etc, also speed up the creation of a new schematic. Over time you will develop your own style using a combination of techniques. To examine in detail and make the schematic fill the screen, select the third icon from the top icon toolbar. This produces a small pointer with a magnifying glass, so place a box around the schematic by holding down the left mouse button **(Lclick)**. The other screen displayed is the management/resource area as shown in Fig. 1.3.

You may have to double left click **(Dlclick)** PAGE1 in the sub directory **DC_circuits.dsn** in order to open up the schematic area where components are placed, although the previous step should avoid having to do this.

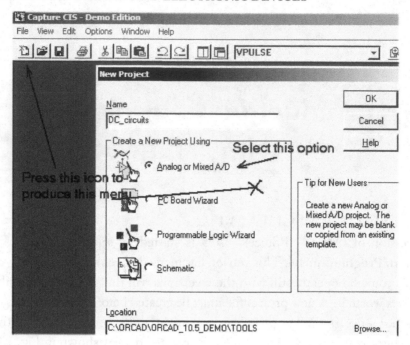

FIGURE 1.1: Creating new project

1.2   LIBRARIES

Libraries have to be added using **Add Library** in Fig. 1.4 by the operator at the be-ginning of simulation. These libraries have the file extension **.olb**, and are located in Tools\Capture\Library\PSpice directory in the main installation directory. The simulation model libraries (**.lib** files) are located under the Tools\PSpice\Library\ directory, under your main installation directory. Selecting **Add Library** opens up a **Browse File** menu where you select one or all of the libraries. From this menu, you may also add customized libraries in-cluding PSpice version 8 libraries. Entering a component name or part of it in the **Part Name** box saves time by not having to search through the libraries. For example, enter the capacitor

FIGURE 1.2: Create a blank project

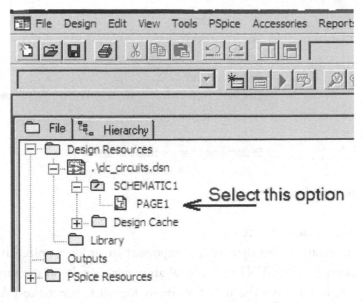

FIGURE 1.3: The management directory

name **c** (PSpice is not case sensitive so capitalized letters are not required), displays the parts in the **ANALOG.olb** library. The wildcard operator * enables a quick search in the library for a component when you are not sure of the complete part name. For example, *741 displays parts with 741 in the name. A DC power supply is called VDC as shown in Fig. 1.4.

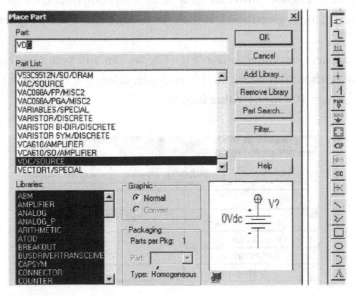

FIGURE 1.4: Analog library

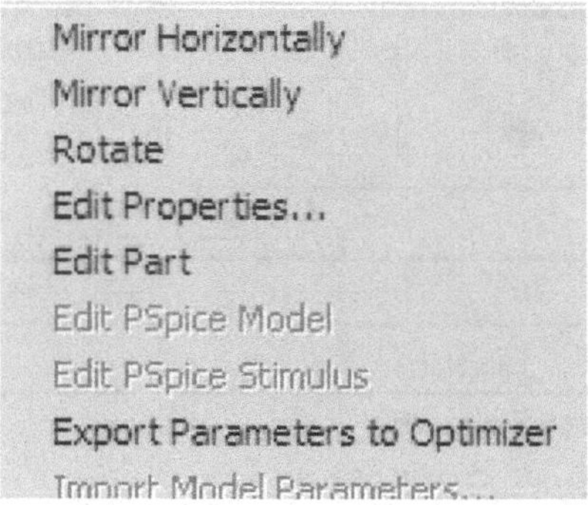

FIGURE 1.5: Select **Edit Properties**

1.2.1   Moving Components Around

It is sometimes necessary to manipulate a component in order to place it with a desired orientation. For example, the uA741 operational amplifier defaults to the noninverting input topmost when placed. To make the inverting input topmost, select the uA741, right click the component and select **Mirror Vertically**. Selecting a component and holding the left mouse button down enables you to move the component. Use the wire icon (**W short-cut**) to connect components. Locate the cursor at one end of the component and press and then release the left mouse button (**Lclick**). To draw a wire segment, **Lclick** the mouse button again and move the mouse from that component end to another component end. Release the left mouse button to change direction. Another technique for connecting components is to select a component, hold the left mouse button and drag to another component, where it should connect. Select a component (turns green) and **Rclick** and select the **Edit Properties** from the list in Fig. 1.5.

In the spreadsheet section shown in Fig. 1.6, we can add new rows and fill in values in the **A** column (more about this later). In this example, select the $R2$ default value of 1 kΩ and replace it by a value e.g. 10k in the **Value** box.

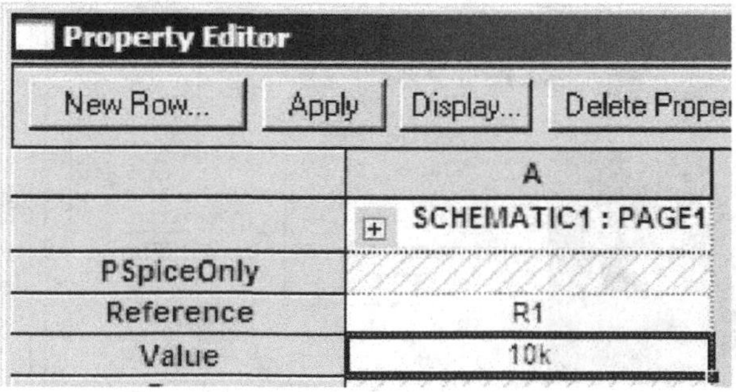

FIGURE 1.6: Changing the part value

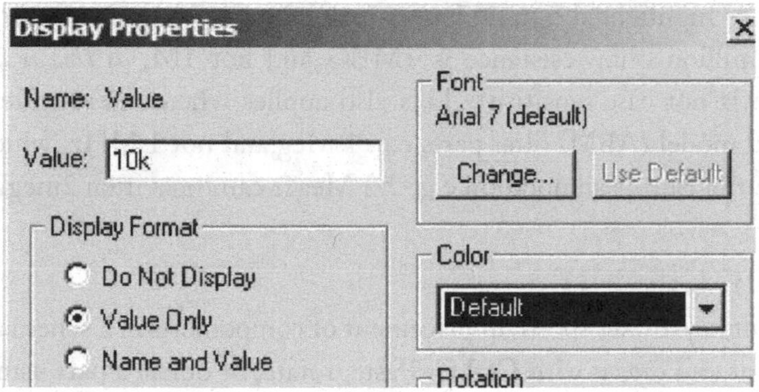

FIGURE 1.7: Changing a component value

1.2.2   Display Properties

Separate the two connected components to a desired position with the left mouse button held
down. To change a part value from the default value, select the default value and the menu in
Fig. 1.7 will appear.

Replace the 1 kΩ default value with a 10k Ω resistor (note there is no space between the 1
and k, and no Ω symbol. For larger resistance values, you may use the exponent system, so that
10k is entered as 1e4. A 10 nF capacitor is 10n (use letter u for micro –the F is optional), and
ten micro henries is 10u (or 10uH). Be careful about capacitor values: A one farad capacitance is
entered as 1, and not 1F, because this would be interpreted as a very small one Femto capacitor.
Table 1.1 shows the symbol, scale, and name for components.

TABLE 1.1: Cpmponent Units

SYMBOL	SCALE	NAME
F	1e−15	Femto-
P	1e−12	Pico-
N	1e−9	Nano-
U	1e−6	Micro-
M	1e−3	Milli-
K	1e+3	Kilo-
Meg	1e+6	Mega-
G	1e+9	Giga-
T	1e+12	Tera-

Those values highlighted in bold letters may cause problems if entered incorrectly. For example, a one million ohm resistance is 1MEG and not **1M**, *as this is a resistance of one milliohm* (PSpice is not case sensitive). This also applies when entering frequency values in analog behavioral model (ABM) filter parts e.g. 1 Meg and not 1 MHz, which is a millihertz. We may also use the European notation e.g. 2.2 MegΩ can be written 2meg2.

1.3   THE DC CIRCUIT

Windows shortcut keystrokes speed up placement of components in a schematic. For example, place a component and *copy* it with **Ctrl C**. Paste, rotate, or delete a part using **Ctrl V, R**, and **Ctrl X**. The first letter of a selected menu is underlined when you press the **Alt** button, so, for example, **Alt W** opens the windows menu. The following are useful shortcut commands:

- To copy a component   **Ctrl** + C
- To paste a component   **Ctrl** + V
- To rotate a component   R
- To discontinue an action   Esc button or right click
- To simulate   **F11** (or use the blue arrow on the top toolbar)
- Wire tool   W

Select the "**get parts**" browser icon (An AND gate symbol on the vertical icon toolbar on the right-hand side) to access the library. Libraries may be added or deleted from this menu. Select the **Place Text** icon **A** from the same toolbar and place your name/date/class group on the schematic. Ohm's law states that the current in a resistance, at a constant temperature, is proportional to the *voltage difference* across the resistance, and inversely proportional to the resistance. To prove Ohm's law (George Simon Ohm 1789–1854), construct the DC circuit shown in Fig. 1.8. Ohm's law states that the current in R_1 is

$$I = \frac{V_1 - V_2}{R_1} = \frac{3 - 1}{1} = 2 \text{ A.} \qquad (1.1)$$

Current direction is set by the relative magnitude and orientation of the two voltage sources. DC current values are displayed on the schematic by selecting the "**I**" and "**V**" icons (but only after simulation). Here, V_1 has a magnitude greater than V_2 so the current flows from V_1 to V_2. However, to plot current in **Probe** for V_1 values, place a current marker from the **PSpice/Markers** menu, or select the icon with the "I" current bubble at the end. *Before any simulation can be carried out, PSpice requires you to set up a new simulation profile.* Indeed, it will not allow you to place markers before this step.

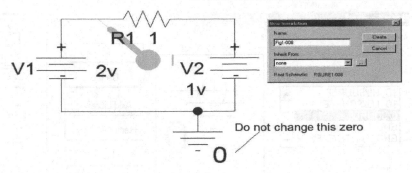

FIGURE 1.8: DC circuit for investigating Ohm's law

1.3.1   New Simulation

In most electronic simulation software programs, it is necessary to create a simulation profile before simulating (see Fig. 1.8). Select the **New Simulation Profile** icon (see the bottom left icon in Fig. 1.10). When you select this and enter none in the **Inherit From** box, it will open up the **Simulation Setting** menu beside it. We may select in the **Inherit From** box, an existing simulation profile with the file extension ***.sim**. This can be handy if we wish to use previous profiles and added libraries. The profile specifies what analysis you want and the parameters necessary for a correct display. This also forces the operator to think about the circuit and the range of values required for a particular design. The evaluation version of 10.5 contains more parts than other versions but it does not, however, allow as many parts to be placed and simulated. Libraries are added when needed and I would suggest that you add the library of customized parts that are required for certain schematics in this, and my other four books on PSpice.

All analog schematics must have a ground symbol otherwise a floating error will be displayed. Ground symbols are placed by selecting, from the icon toolbar on the right, the selection of ground symbols as shown in Fig. 1.9. Select the one shown that has a little zero beside it, **do not delete or change this in any way**. Sometimes this error message is displayed even when you have a ground part, but this error is fixed by placing a large resistance from the offending node to ground. You may select and place any ground symbol, select it and change its name to "zero".

The most commonly used components are: resistance R, inductance L and capacitance C in the **analog.olb** library. Common input signal sources in the **source.olb** library are: AC sources, **VAC**, for frequency response plots and **VSIN** for transient and frequency response plots, VDC (DC supply and step signal source), and ground **GND** (an available icon). Move the mouse pointer over a library to show its location on the hard disk. To examine the schematic in greater detail and make it fill the screen, it is necessary to select the third icon from the

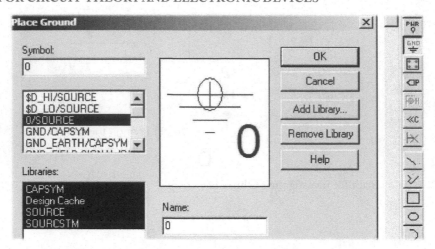

FIGURE 1.9: The earth or ground symbol

left  in the top toolbar. This displays a magnifying glass pointer on the schematic area and is used to place a box around the schematic by holding down the left mouse button.

1.3.2   Main Operational Icons

The main icons are shown in Fig. 1.10. The top toolbar icon shown here is normally located on the right-hand side and is used for creating the schematic. The operational icons shown underneath are for creating simulation profiles, simulating, and placing markers. A new simulation profile is initiated by selecting the first icon. The second icon is pressed when a new profile is created or you wish to change existing parameters. Locating the cursor over an icon will produce an information bubble telling you what it does.

All schematics in this book will be available, so look up the Morganclaypool site at **http://www.morganclaypool.com/**. For example, the first schematic investigates Ohm's law and is named Figure1-008.OPJ (The extra "00" is included to facilitate ordered numerical listing). Hierarchical schematic are examined in book 3 [3], Hierarchical block methods of constructing schematics are useful when the circuit is complex, so consult ref 2, 3 and 6 for further details of this technique.

FIGURE 1.10: Main operational icons

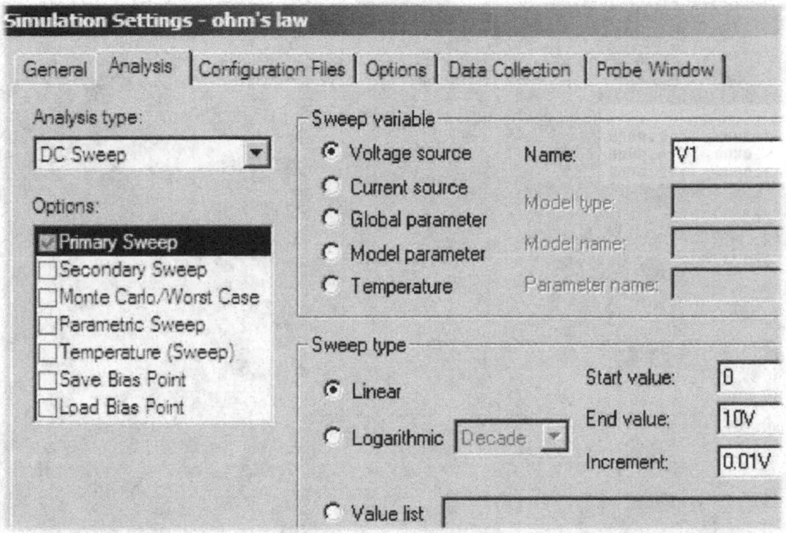

FIGURE 1.11: Sweeping the voltage source V1

1.3.3   Simulation Settings

Set up a new profile as explained before and the **Edit Simulations Settings** menu will open up as in Fig. 1.11. From the **Analysis type** box, select **DC Sweep** and enter the parameters as shown. From the **Sweep variable** section tick **Voltage source** and **Linear** in the **Sweep type** section. Enter the voltage name **V1** in the **Name** box. This voltage source is swept from **Start value:** = 0V to **End value:** = 10V and **Increment:** = 0.01V (*No space* between number and units).

Simulate by selecting the blue triangle icon, or pressing the **F11** key. Fig. 1.12 shows, if no errors, a **Probe** plot of current versus the swept voltage V_1. Incorrect numbers in the set-up parameter, or components not connected together, will produce error messages. Note: If you place the current marker to the right of the resistance, then the graph will have a negative slope so be careful about marker placement. From the Probe output, select the cursor icon and place two cursors on the graph using the left and right mouse buttons. Read the current and voltage differences, and from the inverse of the slope of the graph yields the resistance $R1 = \Delta V/\Delta I$.

Another thing to notice is that the graph does not start at zero but at −1 A. This is because the second DC source sends current in the opposite direction when the first voltage has a zero value.

1.4   POTENTIAL DIVIDER

A voltage transfer function (Alessandro Volta 1745–1827) is the ratio of the output voltage to the input voltage, and is useful for analyzing two-port networks (A port is a pair of terminals,

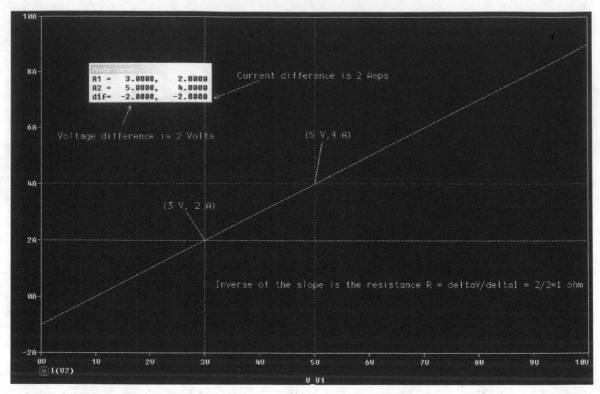

FIGURE 1.12: Current versus swept voltage V1

such as the input or output terminals). A quick way to write a voltage transfer function is to apply the potential divider principle and work out the ratio of the output to input voltages. The potential divider shown in Fig. 1.13 is formed from a voltage V_1 connected across two resistances R_1 and R_2.

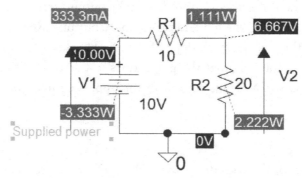

FIGURE 1.13: Displaying DC conditions on a resistive potential divider

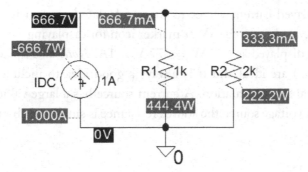

FIGURE 1.14: Current divider network

The transfer function is obtained by substituting the total current $I = V_1/(R_1 + R_2)$ into the output voltage $V_2 = I R_2$.

$$V_2 = \left(\frac{V_1}{R_1 + R_2}\right) R_2 \Rightarrow \frac{V_2}{V_1} = \frac{R_2}{R_1 + R_2}. \tag{1.2}$$

Select the **V I** icons for displaying DC conditions. The new **W** icon shows the power supplied to the circuit and the power dissipated in each resistor after simulation.

1.4.1   Current Divider

The current-divider circuit in Fig. 1.14 shows a current DC source called **IDC** placed in parallel with R_1 and R_2. The current transfer function is a ratio of output and input currents I_{out}/I_{in}.

The voltage across the two resistors is

$$V = I_1 R_1 = I_2 R_2 = I_{DC} R_T. \tag{1.3}$$

I_1 and I_2 are the currents in R_1 and R_2, and R_T is the total resistance. The input current is I_{DC} and the output current is I_2 (Andre Ampère 1775–1836). Substituting the total resistance $R_T = R_1 R_2/(R_1 + R_2)$ into (1.3) gives the current in the second resistor $I_2 = I_{DC} R_T/R_2$, hence the current transfer function is

$$I_2 R_2 = I_{DC} R_T \Rightarrow \frac{I_2}{I_{DC}} = \frac{R_T}{R_2} = \frac{R_1 R_2}{(R_1 + R_2) R_2} = \frac{R_1}{R_1 + R_2}. \tag{1.4}$$

Select the DC current icon and press the **F11** key to simulate. The current in each resistance is

$$I_2 = \frac{R_1 I_{DC}}{R_1 + R_2} = \frac{1k.1}{1k + 2k} A = 0.333 A \tag{1.5}$$

$$I_1 = \frac{R_2 I_{DC}}{R_1 + R_2} = \frac{2k.1}{1k + 2k} A = 0.666 A. \tag{1.6}$$

Compare displayed current values to those calculated in each resistor using equations (1.5) and (1.6). We can also use the Watt marker icon for displaying power on the schematic. The source power is displayed as 677 W = 677 V × 1A. *Note:* all PSpice current and voltage sources (AC and DC) are ideal and it is always a good idea to include some resistances to represent the nonideal source situation. A current source has a large value resistance placed in parallel with it. For a voltage source the source resistance is small and should be placed in series with the voltage.

1.5   EXERCISES

(1) Investigate resistive loading on voltage and current dividers by placing a resistance in parallel with the output resistance, R_2. How does loading change the output voltage/current?

(2) Repeat exercise (1) but investigate loading on the current divider.

CHAPTER 2

The DC Circuit and Kirchhoff's Laws

2.1   MAXIMUM POWER TRANSFER

Maximum power is transferred from a source to an equal-value load and is proved by differentiating the output power with respect to the load resistance and equating to zero. The potential divider in Fig. 2.1 demonstrates maximum power transfer from the source, V source, and source resistance, R source, to a varying load RL.

2.1.1   Param Part

In this example, select the RL default value of 1 kΩ and replace it by a name of your choice e.g. {**Rvar**} in the **Name** box of the **Display Parameter** menu. *The variable resistance name must be enclosed in curly brackets{}*. However, the **Rvar** name has no significance and can be any name you choose. The **PARAM** part, from the **SPECIAL.OLB** library, is very useful for investigating circuit behavior for a range of part values. This part, however, is used and accessed in a different manner to that in previous versions of PSpice. What is required is to place the part using the little AND gate symbol, and when selected it turns green. **Rclick** and select **Edit Properties** to display the spreadsheet shown in Fig. 2.2. In this spreadsheet, we may add new rows and assign values in the **A** column. There is now no limit to the number of parameters you may add in version 10.5, unlike the **PARAM** part in version 8, which was limited to three parameters.

Select **New Row** and enter **Rvar** in the **Name:** box but **NO curly brackets,** and the nominal value of 300 Ω in the **Value** box as shown in Fig. 2.3.

2.1.2   Simulation Settings

Press **Edit Simulation Settings** icon or select the **PSpice/New Simulation Profile** menu. Select **DC Sweep** and enter the parameters shown in Fig. 2.4. **Rvar** is varied **Linear**ly from 1 to 300 Ω in **Increment**s of 1 Ω.

An alternative to the above procedure is to specify a list of values separated by spaces in the **Value list** box.

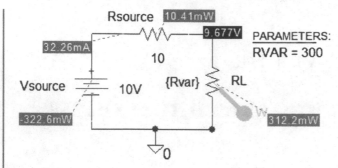

Note if you rotate RL 180 degrees, the power plot I(RL)*V1(RL) will be inverted and it would necessitate placing a minus sign in front of the product to invert.

FIGURE 2.1: Potential divider with variable load

	A
	⊞ FIGURE1-012 : PAGE1 :
PSpiceOnly	TRUE
Reference	1
Value	PARAM
Rvar	300
Location X-Coordinate	560
Location Y-Coordinate	60
Source Part	PARAM.Normal

New Row...   Apply   Display...   Delete Prop

FIGURE 2.2: Edit Properties Param parameters

Add New Row

Name:

Rvar

Value:

300

Enter a name and click Apply or OK to add a column/row to the property editor and optionally the current filter (but not the <Current properties> filter).

No properties will be added to selected objects until you enter a value here or in the newly created cells in the property editor spreadsheet.

☑ Always show this column/row in this filter

Apply   OK   Cancel   Help

FIGURE 2.3: The Param part parameters

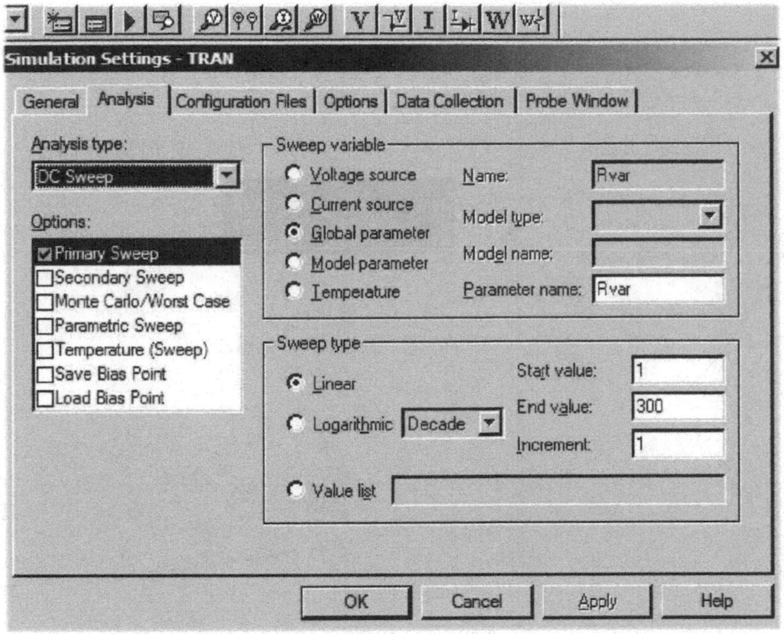

FIGURE 2.4: Analysis and DC sweep

2.1.3 Trace Expression Box

Press **F11** and a blank **Probe** screen appears since no markers were placed on the schematic. To display power dissipated in a device, we may use one of the two techniques. The first technique uses the power marker (**W** icon) that is located in the middle of the device as shown. The second technique is the manual method accessed from the **Probe** screen after simulation when you select the **Trace add** icon ▤ (Or press the **Insert** button on the keyboard). Enter the expression for the load power as **I(R2)*V1(R2)** in the **Trace Expression** box shown in Fig. 2.5.

Note: The resistance orientation has an effect on the final **Probe** display. If the display comes out upside down, then the resistor is the wrong way around and should be rotated around 180° (the **R** key), or place a minus sign in front of the current–voltage product. Component orientation is how PSpice handles conventional current direction flow in R, L, and C, so be careful about the way you place them and marker placement. Fig. 2.6 shows the **Probe** plot when cursors are placed using the left and right mouse buttons. Maximum power is measured

FIGURE 2.5: Adding variables

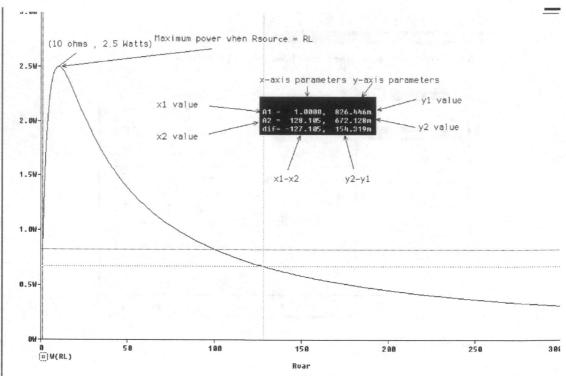

FIGURE 2.6: Power versus Rvar (Linear x-axis variable)

by selecting the maximum **Cursor Peak** ⊼ icons to locate the cursor at the maximum power value. Compare the measured maximum power to the value in (2.1).

$$P_{max} = I^2 R_L = \left[\frac{V_{source}}{R_{source} + R_L}\right]^2 R_L = \left[\frac{V_{source}}{R_L + R_L}\right]^2 R_L = \frac{V_{source}^2}{4R_L} = \frac{100}{4 \times 10} = 2.5 \text{ W}$$

(2.1)

From Probe, select **Plot /Label** to access a menu bar containing many useful display graphic tools such as **Text, line, Arrows, Box**. The plot is copied to clipboard from the Probe/Windows where you have the option of copying with a white or black background.

2.2   CHANGING THE X-AXIS VARIABLE

There are times when you need to change the axis properties such as to change the scale, linear to log, or to change the swept variable. The sub-menu in Fig. 2.7 is accessed by selecting the space beside any x-axis number in **Probe**, or by selecting the **Trace** menu. Tick **Log** to change the x-axis to logarithmic (or select the log icon ▦). A new feature is that we may now place an x-axis name by selecting **Axis Title**. We may change the x- and y-axis range in the **Data Range** box. The x and y grid display can also be changed to the simple display shown in Fig. 2.6. Add

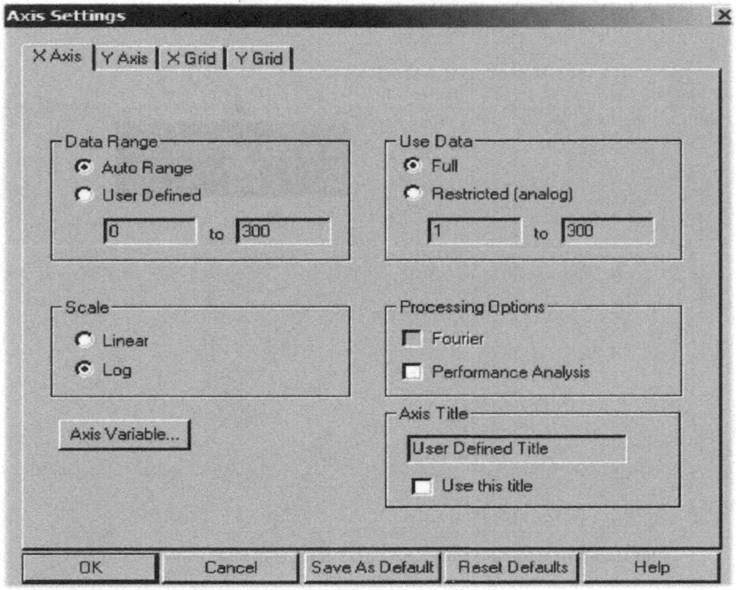

FIGURE 2.7: Changing the x-axis variable

a y-axis title by selecting the y-axis space and entering "Maximum Power Transfer" in the **Axis Title** (a title may be added to the x-axis as well). The Probe plot may be copied to clipboard from the **Windows/Copy to Clipboard** menu where options are offered such as making the background transparent.

2.2.1   The Log Command
All key strokes pressed whilst in the **Probe** environment can be recorded and a file created that can be accessed later from Probe after simulation. This file when run automatically reruns the key pressed sequence to produce the desired Probe screen. This log command file is created by ticking **Log Commands** from the **Probe/File/** menu where you enter a name such as **Figure2-008.cmd** and save. You must un-tick **Log Commands** when you are finished recording all key presses. If you need to re-simulate with different parameters/designs for example, then select **Probe/File/Run Commands** and open **Figure2-008.cmd**. This can be accessed at any time from the probe menu. Fig. 2.8 is not very complicated and hence it is not really necessary here but is very useful when you need to display several plots on different levels. The FFT icon can also be included in this file.

2.3   MESH ANALYSIS
Kirchhoff's voltage law (Gustav Kirchhoff 1824–1887) states that the sum of the potential drops and voltage sources around a closed loop is zero. We may write the mesh voltages' equations

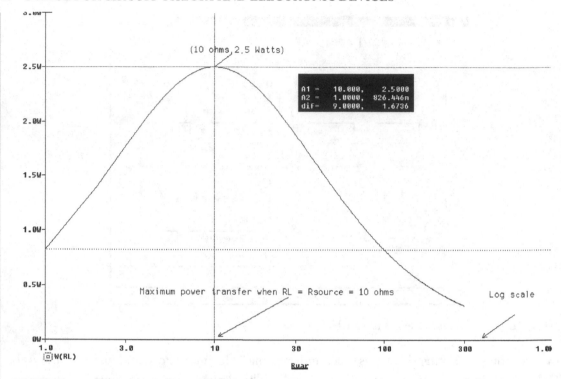

FIGURE 2.8: Power versus Rvar (log *x*-axis variable)

in a simple and safe manner by assigning all mesh currents in the *same clockwise direction*. For example, the Tee network (a popular resistive attenuator) in Fig. 2.9 has two sources E_1 and E_2 (VDC parts) and a *self-impedances in loop one* $R_1 + R_3$ and $R_2 + R_3$ in loop 2. Both impedances are made *positive* and the *mutual impedance* between the loops is $- R_3$.

The sum of the impedances in a loop is called the *self-impedance* and is assigned *a positive sign*, whereas the impedance common to two loops is called the *mutual impedance* and is *always*

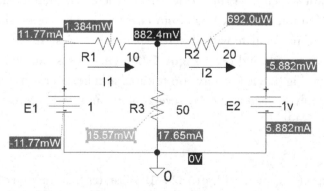

FIGURE 2.9: A Tee network

negative. A negative sign is assigned to each voltage source when the assigned *current* direction is in the *opposite direction* to the rise in a voltage source. Assign all mesh currents in the same direction and applying Kirchhoff's voltage law to yield

$$\text{Loop 1} \qquad E_1 = I_1 R_1 + I_1 R_3 - I_2 R_3 \qquad (2.2)$$

$$\text{Loop 2} \qquad - E_2 = -I_1 R_3 + I_2 R_2 + I_2 R_3. \qquad (2.3)$$

Loop 2 shows a negative sign in front of E_2 because the *current direction* is in the *opposite direction* to the rise in the potential of E_2. Applying these simple rules means we may write the equations in a quick and easy manner, and with less chance of making mistakes. However, (2.2) and (2.3) may be written directly as

$$E_1 = I_1(R_1 + R_3) - I_2 R_3 \qquad (2.4)$$

$$-E_2 = -I_1 R_3 + I_2(R_2 + R_3). \qquad (2.5)$$

The sum of the *self-impedances* in a loop is *positive* and equal to $R_1 + R_3$ in loop 1 and $R_2 + R_3$ in loop 2. The *mutual impedance* is *negative* and equal to $-R_3$ in both equations. The loop equations in matrix form are

$$\begin{bmatrix} E_1 \\ -E_2 \end{bmatrix} = \begin{bmatrix} I_1 \\ I_2 \end{bmatrix} \begin{bmatrix} (R_1 + R_3) & (-R_3) \\ (-R_3) & (R_2 + R_3) \end{bmatrix}. \qquad (2.6)$$

The loop currents are calculated by applying Cramer's rule to (2.6) as

$$I_1 = \frac{\begin{bmatrix} E_1 & (-R_3) \\ -E_2 & (R_2 + R_3) \end{bmatrix}}{\begin{bmatrix} (R_1 + R_3) & (-R_3) \\ (-R_3) & (R_2 + R_3) \end{bmatrix}}$$

$$= \frac{E_1((R_2 + R_3) - (-E_2)((-R_3))}{(R_1 + R_3)(R_2 + R_3) - (-R_3)(-R_3)} = \frac{1 \times 70 - 1 \times 50}{60 \times 70 - 2500} = 11.8 \text{ mA} \qquad (2.7)$$

$$I_2 = \frac{\begin{bmatrix} (R_1 + R_3) & E_1 \\ (-R_3) & -E_2 \end{bmatrix}}{\begin{bmatrix} (R_1 + R_3) & (-R_3) \\ (-R_3) & (R_2 + R_3) \end{bmatrix}}$$

$$= \frac{-E_2((R_2 + R_3) - (E_1)((-R_3))}{1700} = \frac{-1.60 + 1.50}{1700} = \frac{-10}{1700} = 5.9 \text{ mA}. \qquad (2.8)$$

A negative value for the loop current tells you that the current direction is opposite to your assigned direction.

2.4   NODAL ANALYSIS

Nodal analysis is used to solve for the voltage v_x at the junction (node) of R_1, R_2, and R_3 in Fig 2.9. Applying Kirchhoff's current law at the node which states that the sum of the currents at a node is zero:

$$I_1 + I_2 + I_3 = 0. \tag{2.9}$$

We may express each current in terms of the voltages and resistances:

$$\frac{(v_x - V_1)}{R_1} + \frac{(v_x - V_2)}{R_2} + \frac{(v_x - 0)}{R_3} = 0. \tag{2.10}$$

This technique may generate errors in your analysis because of confusion over current and voltage directions. A better technique is to write the node voltage equation using a similar method to that used in mesh analysis. The method is as follows: Show *all* current directions in the *same* direction coming out of a node. This contradicts the normal laws of energy conservation but is done for the reasons discussed in mesh analysis. Name the node under investigation v_x, a voltage with respect to the ground (the reference node). An admittance connected to this node is called a *self-admittance* and is made positive, whereas admittances connected between v_x and any adjacent nodes are called *mutual-admittances* and are made negative. A voltage source attached via admittance to the v_x node is made negative. Thus, we write the nodal equation at the junction of R_1, R_2, and R_3 using this method as

$$v_x \left(\frac{1}{R_1} + \frac{1}{R_2} + \frac{1}{R_3} \right) - \frac{V_1}{R_1} - \frac{V_2}{R_3} = 0. \tag{2.11}$$

Substitute component values and solve for the node voltage v_x as

$$v_x \left(\frac{1}{10} + \frac{1}{20} + \frac{1}{50} \right) - \frac{1}{10} - \frac{1}{20} = 0 \Rightarrow v_x (0.17) = 0.15 \Rightarrow v_x = 0.882 \text{ V}. \tag{2.12}$$

Press **F11** and select the **V** and **I** icons to display current and voltage values on the schematic. The displayed values might be slightly different from the calculated ones because of the number of places of decimals or **<u>Significant Digits</u>**. However, you may change the number of displayed digits by selecting the menu sequence in Fig. 2.10.

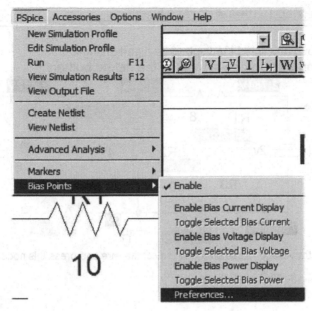

FIGURE 2.10: Changing the digits

Select the **PSpice/Bias Points/Preferences** menu. Enter 4 in the **Displayed Precision** as shown in Fig. 2.11 and it will display the number to four decimal points.

After simulation, select the **V/I** icons to display DC current and voltages shown in Fig. 2.12.

To remove redundant voltage or current displays, press the icon to the right of the voltage or current icons. To add extra voltage displays on a wire, select the wire and press the little node junction icon beside the main **V** icon. Make sure **Bias Point** is ticked in the **Analysis Setup** menu.

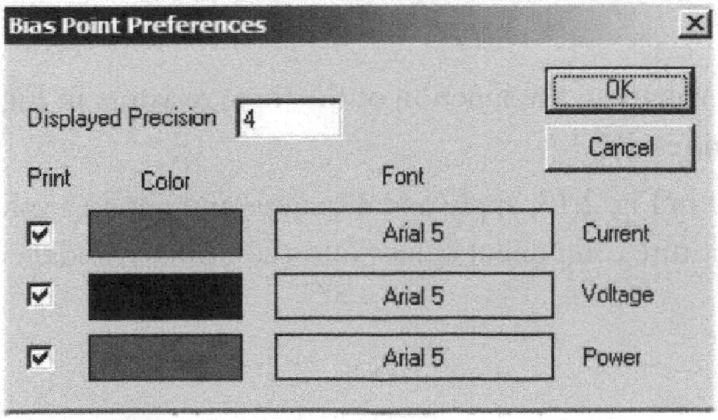

FIGURE 2.11: Setting the digits

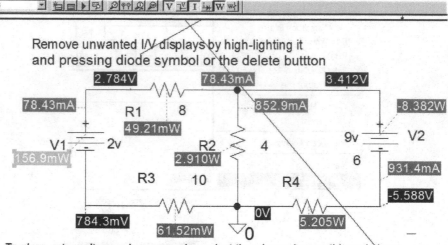

FIGURE 2.12: Displaying DC conditions

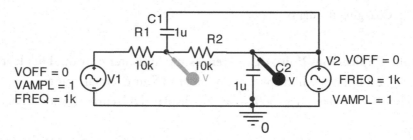

FIGURE 2.13: Part of a Sallen and Key active filter

2.5 EXERCISES

(1) In Fig. 2.1, vary the source resistance and investigate maximum power transfer. (Hint: An ideal voltage has a zero resistance source hence all power is absorbed by the load.)

(2) Calculate the voltage at the junction of the three resistors in Fig. 2.12 and compare to the simulation results.

(3) For the circuit in Fig. 2.13, apply nodal analysis and obtain a value for the node voltage v. Hint: Substitute component values into the primary nodal equation before solving for v.

CHAPTER 3

Transient Circuits and Laplace Transforms

3.1 TRANSIENT ANALYSIS

Complex signals, such as step or pulse-type signals applied to CR, LR circuits, and LCR circuits, are analyzed in a much simpler fashion using the Laplace transform (Pierre Laplace 1749–1825). Laplace tables and partial fraction expansion methods are used when transforming from the s-domain back to the time domain.

3.2 LAPLACE TRANSFORM AND CAPACITANCE

The Laplace Transform of a function $f(t)$ is

$$L_T\{f(t)\} = F(s) = \int_0^\infty f(t)e^{-st}dt \tag{3.1}$$

The current–voltage relationship for a capacitor in the time domain, with an initial voltage across the capacitor plates V_0 (consider this as a step voltage), is

$$v_C(t) = \frac{1}{C}\int_0^t i(t)dt + V_0 \tag{3.2}$$

Laplace transforming equation (3.2):

$$L_T\{v_C(t)\} = L_T\left\{\frac{1}{C}\int_0^t i(t)dt + V_0\right\} \tag{3.3}$$

The Laplace transform for integration and step functions, is 1/s, hence (3.3) becomes

$$V_c(s) = \frac{1}{sC}I_c(s) + \frac{V_0}{s} \tag{3.4}$$

All transformed circuit variables are shown capitalized in the Laplace equivalent circuit. Figs. 3.1(a) and 3.1(b) show a capacitor representation in the time and s domains where the reactance $1/sC$ I is in series with the initial condition voltage, V_0/s, but acting in the sense shown in Fig. 3.1(b).

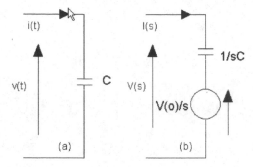

FIGURE 3.1: Transformed capacitance

3.3   INDUCTANCE

The current–voltage–time relationship for an inductor, with an initial current, I_0, is

$$v_L(t) = L\frac{di}{dt} + I_0 \qquad (3.5)$$

The Laplace transform of (3.5) is

$$L_T\{v_L(t)\} = V_L(s) = L_T\left(L\frac{di}{dt} + I_0\right) = s\,L I(s) - L I_0 \qquad (3.6)$$

An inductor representation in the time domain is shown in Fig. 3.2(a) and the Laplace transform of an inductor is shown in Fig. 3.2(b) as a reactance sL ohms in series with the initial condition voltage, LI Volts but *in the same direction as the input voltage.*

The transformed circuit may also be expressed as a Norton equivalent circuit with the Norton admittance in parallel with a current source.

3.4   FIRST-ORDER CR AND LR CIRCUITS

Figure 3.3 shows a high-pass filter that is used to couple amplifier stages together for the purposes of DC isolating each stage. A pair of differential markers measures the voltage across

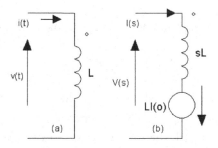

FIGURE 3.2: Transformed inductance

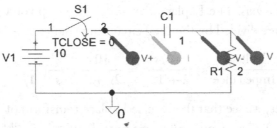

FIGURE 3.3: High-pass CR circuit

a component that is not connected to the ground reference point, such as the voltage across the capacitor. Select the **PSpice/Markers/ Voltage Differential** to place two consecutive markers, where one marker has a plus symbol and the other has a negative symbol (or use the icons from the top marker icon toolbar).

Apply the Laplace transform and obtain expressions for the capacitor and resistor voltages. We may obtain current or voltage expressions in time using the inverse Laplace transform table given in the appendix. After simulation, compare your calculations to the simulation results for currents and voltages evaluated at certain times. The **Sw_tClose** switch part from the **eval.olb** library is closed at time $t = 0$ s (Assume the capacitor is initially uncharged). Evaluate the capacitor voltage at $t = 5$ s, where $V1 = 10$ V, $R_1 = 2\,\Omega$, $C = 1$ F. Capacitance entered in PSpice as 1 F (no space between the number and the unit) is interpreted as 1 Femto farads—a much smaller capacitor, *so leave out the* **F** *symbol*.

3.4.1   Solution

The first switch, **sw_tClose,** closes at a certain time (Note: the **sw_tOpen** switch opens at a specified time). **DLclick** the **sw_tClose** part to set the following parameters:

Time to close	= TCLOSE = 0 s,
Transient time	= TTRAN = 1 us,
Switch resistance when closed	= RCLOSED = 0.01 Ω, and
Switch resistance when opened	= ROPEN = 1 Meg

Other mechanical switches are available in the **discrete.olb** library. The voltage across the capacitor increases exponentially because the current into the capacitor accumulates charge on the capacitor plates. The capacitor current, however, decreases exponentially with time because the capacitor voltage that is building up with time opposes the source voltage. The circuit time constant, $\tau = CR$, is the time it takes for the capacitor voltage to charge up to 63 % of the applied step voltage. The time constant is also found by extrapolating the initial slope of the capacitor voltage with a straight line until it intersects a line drawn across from the final capacitor voltage to the y-axis. The time constant τ is then measured by dropping a perpendicular line from the intercept to the time axis. The Laplace transform for a step voltage V_1 is V_1/s, hence the current is the step voltage divided by the total circuit impedance $R + 1/sC$ as

$$I_R(s) = \frac{\text{Voltage}}{\text{Impedance}} = \frac{10/s}{2 + 1/s} = \frac{10}{2s + 1} = \frac{5}{s + 1/2} = \frac{5}{s + 0.5} \qquad (3.7)$$

From the Laplace tables, we see that the inverse Laplace transform of $1/(s + a)$ is e^{-at}, where $a = 0.5$ so that Eq. (3.7) when converted to the time domain is $5e^{-0.5t}$. The current at $t = 5$ s is

$$i_R(t) = 5e^{-0.5t}\Big|_{t=5} \Rightarrow i_R(5) = 5e^{-0.5 \times 5} = 5e^{-2.5} = 0.414 \text{ A} \qquad (3.8)$$

3.4.2 Partial Fraction Expansion

Partial fraction expansion (PFE) is necessary to transform an equation in s, back to an equation in time, where the equation in s has no direct equivalent in the Laplace tables. For example, the capacitor voltage in the previous example is obtained by multiplying the current in (3.8) by the capacitor reactance to yield

$$V_c(s) = I(s)\left(\frac{1}{sC}\right) = \left(\frac{5}{s + 0.5}\right)\left(\frac{1}{s}\right) = 5\frac{1}{(s + 0.5)(s)} \qquad (3.9)$$

The two functions $1/(s + 0.5)$ and $1/s$ are in the Laplace tables, but the product of the two is not, so we need to separate the parts using the partial fraction expansion method. The constant 5 is temporarily ignored, so separate the two parts in (3.9) by introducing two new constants A and B as

$$\frac{1}{(s + 0.5)(s)} = \frac{A}{(s + 0.5)} + \frac{B}{(s)} = \frac{A(s) + B(s + 0.5)}{(s + 0.5)(s)} = \frac{(A + B)s + B0.5}{(s + 0.5)(s)} \qquad (3.10)$$

Combine all s terms and equate the right and left terms of the top part of (3.10) i.e.

$$1 = (A + B)(s) + B0.5 \Rightarrow (A + B)(s) = 0 \text{ and } 0.5B = 1 \Rightarrow B = 2 \qquad (3.11)$$

Now $A + B = 0$ since s is not zero and there are no s terms on the LHS of (3.11). This implies that $A = -B = -2$. Substitute these values back into (3.10):

$$V_c(s) = \left(\frac{5}{s + 0.5}\right)\left(\frac{1}{s}\right) = 5\frac{1}{(s + 0.5)(s)} = 5\left[\frac{2}{s} - \frac{2}{s + 0.5}\right] = 10\left[\frac{1}{s} - \frac{1}{s + 0.5}\right] \qquad (3.12)$$

The inverse Laplace equivalent for the two parts of (3.12) yields the capacitor voltage in the time domain as

$$v_c(t) = 10(1 - e^{-0.5t}) \; V \qquad (3.13)$$

The capacitor voltage at time $t = 5$ s is

$$v_c(5) = 10(1 - e^{-0.5\times5}) = 10(1 - e^{-2.5}) = 9.17 \text{ V} \qquad (3.14)$$

3.4.3 Initial Conditions

There are several methods for setting initial circuit conditions. The **Run to time** should be set to at least five time constants in order to display an almost-charged condition on the capacitor. The first method is to set **Analysis type: Time Domain (Transient), Run to time** = 15 s, **Maximum step size** = 20 ms and tick **S̲kip initial transient solution**. This allows PSpice to charge the capacitor from a *zero charge state*. The second technique is to select the capacitor and type in zero volts on an **IC** parameter (a capacitor parameter). The third technique uses the **IC1** part placed at a junction or the **IC2** part placed across the capacitor and a desired voltage entered. These parts selected from the **special.olb** library are used to place a final voltage on a capacitor thus speeding up charging and, hence, simulation times. Press **F11** to simulate and display the voltages across C and R as shown in Fig. 3.4.

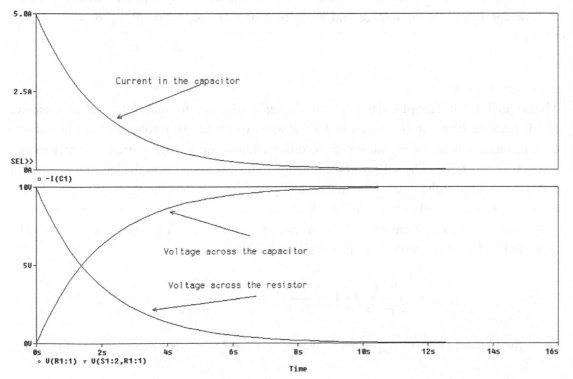

FIGURE 3.4: Transient response

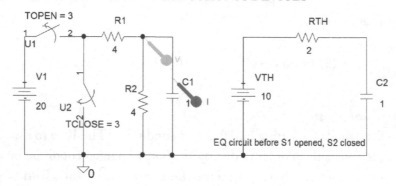

FIGURE 3.5: The original and Thévenin equivalent circuits

3.5 EXAMPLE 2

Fig. 3.5 shows an electrical circuit and the Thévenin equivalent circuit. With C initially uncharged at time $t = 0$ s, and switch S1 (**Sw_tOpen** part) closed, the capacitor starts to charge. Switch S2 (**Sw_tClose** part) is then closed, and S1 opened simultaneously after 3 s. Obtain an expression, using the Laplace Transform, for the current in C as a function of s (Hint: Apply Thévenin's theorem to the left of C, treating C as the load). Obtain an expression in time for the capacitor voltage and evaluate it at $t = 3$ s. $V1 = 20$ V, $R_1 = R_2 = 4\,\Omega$, $C = 1$ F.

3.5.1 Solution

The original circuit is simplified by applying Thévenin's theorem to yield an equivalent resistance of 2 Ω because the resistors are now in parallel when you replace the voltage source by a short-circuit link. In this example, the source is considered an ideal voltage source, and hence has a zero source resistance. The voltage at the junction of the two resistors to the left of the capacitor is determined by applying the potential divider principle. Remove the load C. The Thévenin equivalent voltage is $(10/s)$ V, i.e. half the input voltage (look up potential divider in the index). The complete equivalent circuit is shown at the right in Fig. 3.5. The current is calculated by dividing the Thévenin step voltage by the circuit impedance as

$$I_R(s) = \frac{10/s}{2 + 1/s}\left(\frac{s}{s}\right) = \frac{10}{2s + 1} \div \frac{2}{2} = \frac{5}{s + 1/2} = \frac{5}{s + 0.5}$$

From the Laplace tables in Appendix A, we convert $I_R(s)$ to current as a function of time

$$I_R(t) = 5e^{-0.5t}$$

The current at 3 s is $I_R(3) = 5e^{-0.5 \times 3} = 5e^{-1.5} = 1.11$ A. An expression for the capacitor voltage is obtained by multiplying the current by the capacitor reactance as

$$V_c(s) = I_R(s)\left(\frac{1}{sC}\right) = \left(\frac{5}{s+0.5}\right)\left(\frac{1}{s}\right)$$

There are no entries in the Laplace tables for this product, so we need to apply PFE in order to separate the voltage expression into two functions that are in the tables. For example:

$$5\frac{1}{(s+0.5)(s)} = 5\left(\frac{A}{s+0.5} + \frac{B}{s}\right) = 5\left(\frac{A(s) + B(s+0.5)}{(s+0.5)(s)}\right) = 5\left(\frac{(A+B)s + B0.5)}{(s+0.5)(s)}\right)$$

To determine A and B, equate the top part of the left-hand side of this equation to the RHS:

$$1 = (A+B)(s) + B0.5 \Rightarrow (A+B)(s) = 0 \Rightarrow (A+B) = 0 \Rightarrow A = -B$$

This means that $1 = 0.5B \Rightarrow B = 2$

$$Vc(s) = I(s)\left(\frac{1}{sC}\right) = \left(\frac{5}{s+0.5}\right)\left(\frac{1}{s}\right) = 5\left[\frac{2}{s} - \frac{2}{s+0.5}\right] = 10\left[\frac{1}{s} - \frac{1}{s+0.5}\right]$$

From the Laplace tables, we see that $1/s$ is equivalent to 1 in the time domain and $1/(s+0.5)$ is equivalent to $e^{-0.5t}$. The capacitor voltage in the time domain is

$$v_c(t) = 10(1 - e^{-0.5t})\ V$$

At $t = 3$ s, the 20 V supply is disconnected by opening S1 and closing S2, thus allowing the capacitor to charge to $v_c(3) = 10(1 - e^{-0.5 \times 3}) = 7.76$ V. The circuit now comprises a voltage source $7.76/s$ V in series with a 2 Ω resistance and a 1 F capacitor. However, this voltage source is not a constant voltage supply and so decays exponentially $v_c(t) = 10e^{-0.5t}$. The current changes direction instantaneously because the capacitor cannot charge instantaneously but decays in an exponential manner. Select the **Analysis/Transient** menu and set **Run to time** to 15 s and **Maximum step size** (default value). Press **F11** to simulate and produce the current and voltage waveforms shown in Fig. 3.6.

3.6   EXAMPLE 3

Fig. 3.7 shows a switch open for a long time which is then closed at $t = 0.2$ s. Obtain an expression for the inductor current as a function of s. Using the Laplace tables, get an expression for the inductor current in time and evaluate at $t = 0, 0.2$ and 0.3 s. Assume zero initial conditions and $R_1 = R_2 = 5\ \Omega$, $L = 2$ H, $V_1 = 1$ V.

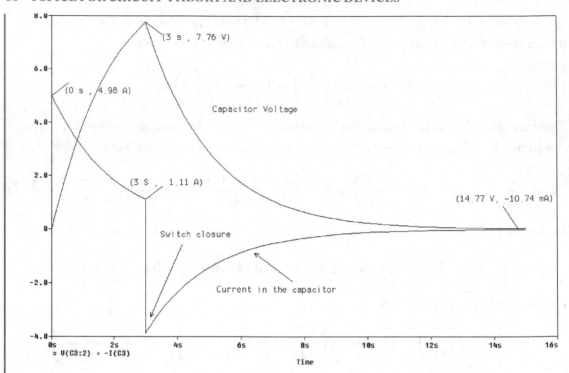

FIGURE 3.6: Current and voltage waveforms

3.6.1   Solution

Laplace transforming the circuit yields the current as

$$I_L(s) = \frac{1/s}{2s + 10} = \frac{1}{s}\frac{1}{2s + 10} = \frac{1}{(s)}\frac{1/2}{(s + 5)} = 0.5\left[\frac{1}{(s)(s + 5)}\right] = \frac{A}{s} + \frac{B}{s + 5}$$

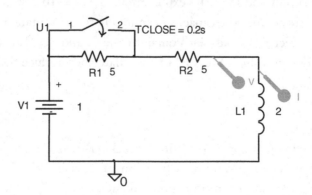

FIGURE 3.7: Switch closed at 0.2 s

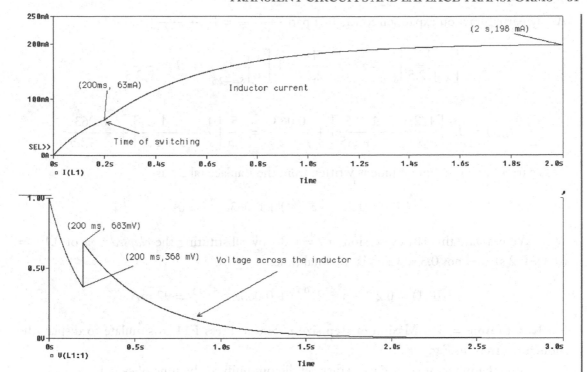

FIGURE 3.8: Inductor current

Ignore the 0.5 factor and equate the top left-hand side of the equation to the right-hand side as

$$1 = (A+B)(s) + A5 \Rightarrow A = 1/5 \text{ and } (A+B)(s) = 0 \Rightarrow (A+B) = 0, \text{ or, } B = -A = -1/5$$

Another method of solving PFE coefficients is $[\frac{1}{s+5}]_{S=0} \Rightarrow A = \frac{1}{5}$ and $[\frac{1}{s}]_{S=-5} \Rightarrow B = -\frac{1}{5}$

$$I_L(s) = 0.5 \left[\frac{1/5}{s} - \frac{1/5}{s+5} \right] = \frac{0.5}{5} \left[\frac{1}{s} - \frac{1}{s+5} \right] = 0.1(1 - e^{-5t})$$

The current at $t = 0.2$ s is $i = 63$ mA (see Fig. 3.8). Closing the switch at $t = 0.2$ s reduces the circuit resistance to 5 Ω, However, the inductor initial voltage is now equal to the current at 0.2 s multiplied by the inductance, i.e. $VL = I(0.2)L = (63 \text{ mA})(2 \text{ Henries}) = 0.106$ V. The initial voltage direction is in the opposite direction to the voltage drop across the inductance. The current is the voltage $(1/s + 0.106)$/impedance i.e.

$$I_L(s) = \frac{1/s + 0.126}{2s + 5} = \frac{1/2s}{s + 2.5} + \frac{0.063}{s + 2.5} = \frac{0.5}{(s)(s+2.5)} + \frac{0.063}{s + 2.5}$$

Apply partial fraction expansion to the first part $\frac{1}{(s)(s+2.5)} = [\frac{A}{s} + \frac{B}{s+2.5}]$ and solve

$$\left[\frac{1}{s+2.5}\right]_{S=0} \Rightarrow A = \frac{1}{2.5} \text{ and } \left[\frac{1}{s}\right]_{S=-2.5} \Rightarrow B = -\frac{1}{2.5}$$

$$I_L(s) = 0.5\left[\frac{1/2.5}{s} - \frac{1/2.5}{s+2.5}\right] + \frac{0.063}{s+2.5} = \frac{.5}{2.5}\left[\frac{1}{s} - \frac{1}{s+2.5}\right] + \frac{0.063}{s+2.5}$$

This current as a function of time is written from the Laplace tables as

$$I_L(t) = 0.2(1 - e^{-2.5t}) + 0.063e^{-2.5t} \text{ mA}$$

Note: We evaluate this last expression at $t = 0.3$ s by substituting the *elapsed time* of 0.1 s = (0.3 s–0.2 s), and not 0.3 s, which is the total time.

$$I_L(0.1) = 0.2(1 - e^{-2.5\times0.1}) + 0.063e^{-2.5\times0.1} = 92 \text{ mA}$$

Set **Run to time** = 3 s, **Maximum step size** = 500 u. Press **F11** to simulate to display the inductor current in Fig. 3.8.

Note the inductor voltage has a distinct discontinuity at the time of switching.

3.7 EXAMPLE 4

The switch in Fig. 3.9 is initially opened and then closed at $t = 80$ ms. Using the Laplace transform, obtain an expression in s for the inductor current. Hence, calculate a value for the inductor current at $t = 80$ ms. Assume the inductor is initially uncharged. $R_1 = 5\ \Omega$, $R_2 = 20\ \Omega$, $L = 0.1$ H, $I_1 = 2$ A, $I_2 = 3$ A.

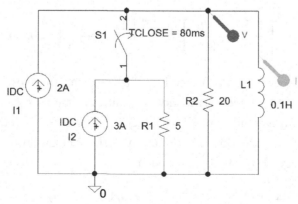

FIGURE 3.9: Problem 4 using parallel current sources

3.7.1   Solution

Before the switch is closed, calculate the inductor current by applying the current divider principle to I_1, R_2, and L_1:

$$I_L(s) = \frac{2}{s}\frac{20}{20+s0.1} = \left(\frac{2}{s}\right)\left(\frac{200}{200+s}\right) = \frac{400}{s}\frac{1}{(200+s)} = 400\left[\frac{A}{s} + \frac{B}{s+200}\right]$$

Solve for A and B using partial fraction expansion $[\frac{1}{s+200}]_{s=0} \Rightarrow A = \frac{1}{200}$ and $[\frac{1}{s}]_{s=-200} \Rightarrow B = -\frac{1}{200}$.

$$I_L(s) = \frac{400}{200}\left[1 - \frac{1}{s+200}\right] \Rightarrow I_L(t) = 2\left[1 - e^{-200t}\right] \ A$$

When the switch is closed at $t = 80$ ms, the inductor current is $I_L(80 \text{ ms}) = 2(1 - e^{-200 \times 80\text{ms}}) = 2$ A. The initial inductor current I_0 produces a back EMF $I_0 = 80$ mV. The equivalent circuit is a generator in series with the inductor reactance. Complete the solution by combining the two current sources into one source and the two resistances into one resistance. Set Analysis to **Analysis type: Time Domain (Transient), Run to time** = 500 ms, and **Maximum step size** = 10 us. Press **F11** to plot the inductor current shown in Fig. 3.10.

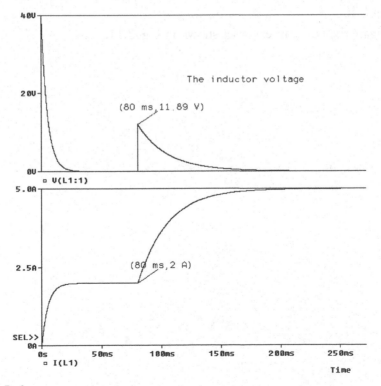

FIGURE 3.10: Inductor current

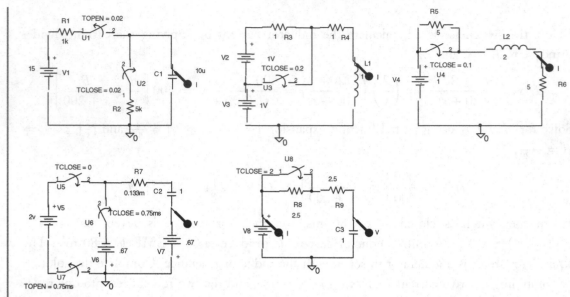

FIGURE 3.11: Switching circuits

3.8   EXERCISE

(1) Investigate the transient exercises shown in Fig. 3.11.

CHAPTER 4

Transfer Functions and System Parameters

4.1 TRANSFER FUNCTIONS

A transfer function comprises numerator and denominator parts with each part containing one or more factors. We may use a **LAPLACE** part to plot the transfer function frequency response or the impulse response. The default Laplace part has a unity numerator and denominator is $1 + s$, which is a low-pass filter with a 1 rs^{-1} cut-off frequency. Each transfer function factor entered is separated by a multiplier operator "*". For example, if the numerator is 5 s, then enter it as 5*s. A fourth-order polynomial, factored into two second-order polynomials, has each polynomial enclosed by brackets () but separated by a "*" symbol between each pair of brackets.

4.2 BUTTERWORTH TRANSFER FUNCTIONS AND THE LAPLACE PART

A second-order transfer function is formed from an inverted second-order Butterworth loss function with a cut-off frequency $\omega_0^2 = 100 \Rightarrow \omega_0 = 10 \text{ rs}^{-1} \Rightarrow f_0 = 10 \text{ rs}^{-1}/(2\pi) = 1.59$ Hz. The Butterworth loss function is de-normalized by replacing the normalized complex frequency variable $ with the complex frequency variable, s, divided by the cut-off frequency [ref: 1]. The transfer function is therefore

$$H(s) = \frac{V_{\text{out}}}{V_{\text{in}}} = \frac{1}{\$^2 + 1.414\$ + 1}\bigg|_{\$=s/10} = \frac{1}{(\frac{s}{10})^2 + 1.414\frac{s}{10} + 1} = \frac{100}{s^2 + 14.14s + 100}$$

(4.1)

Draw the schematic in Fig. 4.1 and use the **Net Alias** icon to place a name on the output wire.

Select the **Laplace** part, **Rclick** and select **Edit Properties**. In the **DENOM**inator box, enter the transfer function denominator $(s*s + 14.14*s + 100)$, and 100 in the **NUM**erator box. In the top right-hand corner press the small x to exit the properties. To plot the frequency response, place a dB marker (From the **PSpice/Markers/Advanced** menu) on the output port. From the **Analysis Setup** menu, set the sweep parameters shown in Fig. 4.2 to plot 1001 points for each decade in the response from 0.1 Hz to 10 kHz.

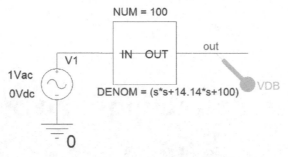

FIGURE 4.1: Laplace part for evaluating step and frequency responses

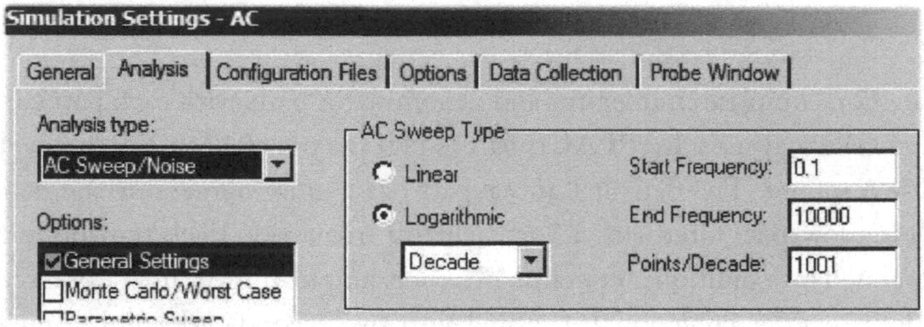

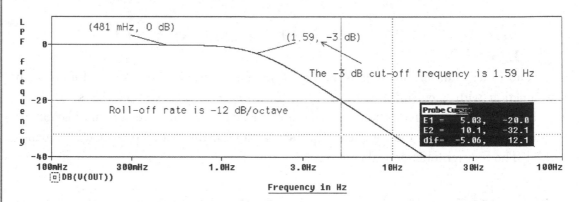

FIGURE 4.2: Analysis Setup parameters

FIGURE 4.3: Frequency response

Simulate, and if no errors are present then the frequency response in Fig. 4.3 should appear. The frequency at which the output is down by −3 dB is measured using the cursor. Make sure it agrees with your calculations.

Place a phase marker from the **PSpice/Markers/Advanced** menu on the output node, and simulate. Construct the circuit for obtaining the step response and set the transient analysis parameters as shown in Fig. 4.4.

The input signal for producing the transient response is now considered.

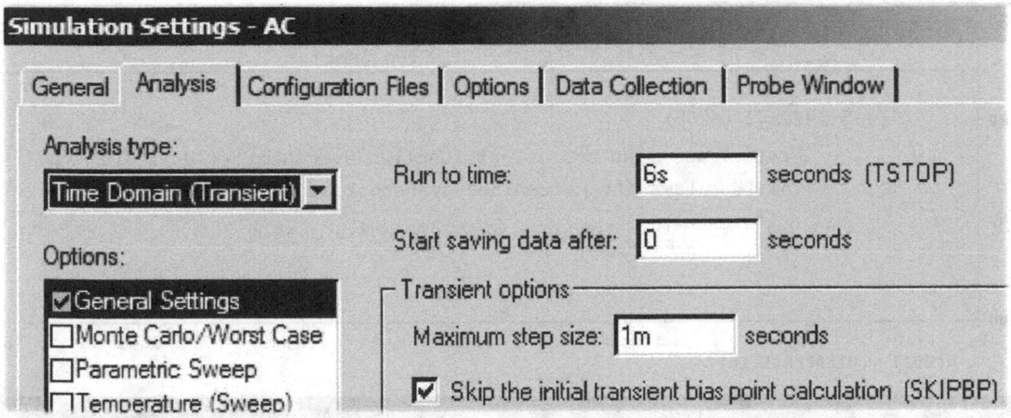

Simulation Settings - AC

General | **Analysis** | Configuration Files | Options | Data Collection | Probe Window

Analysis type:

Time Domain (Transient) ▼

Run to time: 6s seconds (TSTOP)

Start saving data after: 0 seconds

Options:

☑ General Settings
☐ Monte Carlo/Worst Case
☐ Parametric Sweep
☐ Temperature (Sweep)

┌ Transient options ─────────────

Maximum step size: 1m seconds

☑ Skip the initial transient bias point calculation (SKIPBP)

FIGURE 4.4: Transient analysis parameters

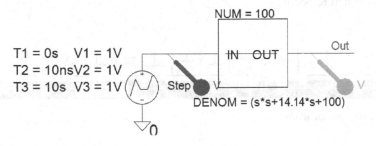

FIGURE 4.5: Step response testing

4.2.1   Piece-wise Linear Part (VPWL)

A piece-wise linear (**VPWL**) shown in Fig. 4.5, part generates a unit step function whose parameters are displayed on the schematic. Alternatively, use a VDC part as a step voltage source.

The step response in Fig. 4.6 shows the step response as an exponentially decaying sinusoidal signal. Separate the input and output signals using **Alt PP, Ctrl X, Ctrl V** on a variable selected at the bottom left of the **Probe** output.

4.3   PROBE GRID AND CURSORS ICONS

The function of each cursor icon is given in a little information box that pops up when you place the mouse cursor on each one as shown in Fig. 4.7. To place specific values on a plot, turn on the cursor, select the maximum-value icon (fifth one from the left) to place intersecting cursors at the maximum value.

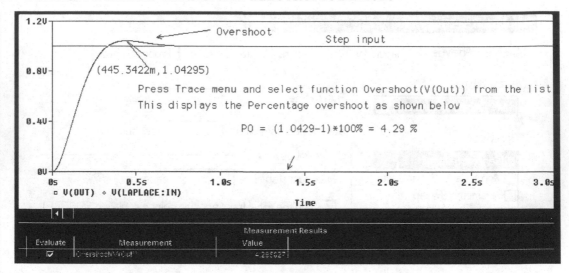

FIGURE 4.6: Unit step response

FIGURE 4.7: Probe cursor icons

Pressing the second last icon on the right places a pair of x and y values at the largest value location. This may be repeated for other peaks by pressing the maximum value icon again and pressing the second last icon, where it places another pair of x and y values at the second largest value, and so on.

The percentage overshoot (PO), is the maximum value minus the step value divided by the step value and multiplied by 100 %. For a unit step, the overshoot is the maximum value of the step response minus one and then multiplied by 100 %. So, in this case it is (1.0429-1)*100 % = 4.29 % (see Section 4.9).

4.4   ELAPLACE PART AND THE STEP RESPONSE

We can measure system parameters such as rise time and steady-state values by applying a step signal and measuring different parameters from the step response. In second-order systems, we measure overshoot, settling time and natural frequency from the step response. The **ELAPLACE** part in Fig. 4.8 can be used to obtain step and impulse responses by entering the transfer function into the **XFORM** box.

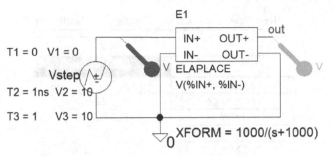

FIGURE 4.8: The Elaplace part

	A
	⊞ FIGURE1-045 : PAGE1 :
PSpiceOnly	TRUE
Reference	Vstep
Value	VPWL
AC	0
BiasValue Power	0W
DC	0
Location X-Coordinate	500
Location Y-Coordinate	345
Schematics' Source Library	C:\MSimEv_8\lib\SOURCE.slb
Source Part	VPWL.Normal
T1	0
T2	1ns
V1	0
V2	10
V3	10

Buttons: New Row... | Apply | Display... | Delete Property

FIGURE 4.9: VPWL step generator spreadsheet

A low-pass filter transfer function, with a cut-off frequency 1000 r/s ($f_c = 159$ Hz), is expressed as

$$H(s) = \frac{V_{\text{out}}}{V_{\text{in}}} = \frac{\omega_p}{s + \omega_p}\bigg|_{\omega=1000} = \frac{1000}{s + 1000} \qquad (4.2)$$

Select the **VPWL** generator part, **Rclick** and select **Edit Properties** and enter the parameter as shown in Fig. 4.9. This creates a step function with the voltage, at time $t_1 = 0$ s, equal to 10 V.

Rclick the **ELAPLACE** and select **Edit Properties** to display the part properties shown in Fig. 4.10. Enter the transfer function **100/(s+100)** in the **A** column of the **XFORM** row.

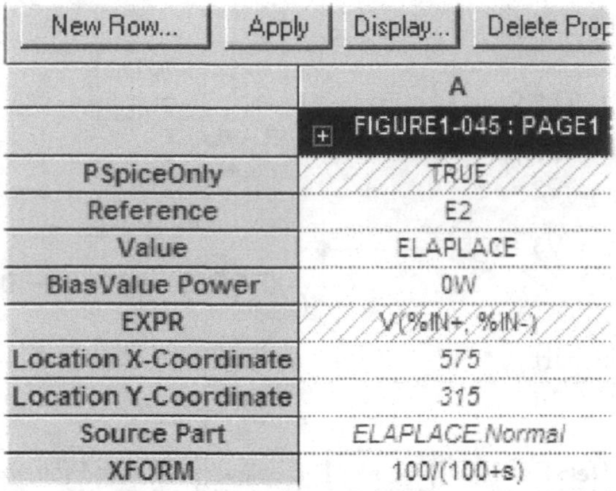

	A
New Row...	Apply	Display...	Delete Prop
	FIGURE1-045 : PAGE1
PSpiceOnly	TRUE
Reference	E2
Value	ELAPLACE
BiasValue Power	0W
EXPR	V(%IN+, %IN-)
Location X-Coordinate	575
Location Y-Coordinate	315
Source Part	ELAPLACE.Normal
XFORM	100/(100+s)

FIGURE 4.10: Elaplace part parameter spreadsheet

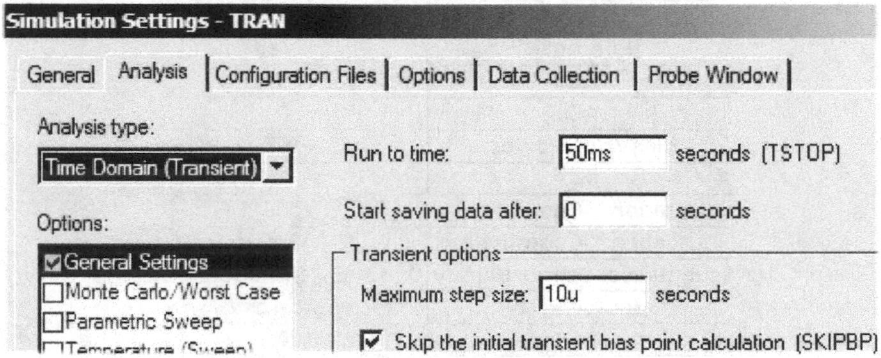

Simulation Settings - TRAN

General | Analysis | Configuration Files | Options | Data Collection | Probe Window

Analysis type:
[Time Domain (Transient) ▼]

Run to time: [50ms] seconds (TSTOP)

Start saving data after: [0] seconds

Options:
☑ General Settings
☐ Monte Carlo/Worst Case
☐ Parametric Sweep
☐ Temperature (Sweep)

Transient options
Maximum step size: [10u] seconds

☑ Skip the initial transient bias point calculation (SKIPBP)

FIGURE 4.11: Transient Analysis parameters

Terminate the **ELAPLACE** output with a port left part from the right-hand menu and set the **Transient Analysis** parameters as shown in Fig. 4.11.

Press the simulate icon (Or the **F11** key), and the step response in Fig. 4.12 should appear.

We will show in Section 4.9 how to measure the rise time.

4.5 CHEBYCHEV TRANSFER FUNCTIONS IMPULSE RESPONSE

A second-order Chebychev loss function with 2 dB ripple ($\varepsilon = 0.746$) in the passband is $\$^2 + 0.804\$ + 0.823$. Invert this function and denormalize by replacing $\$$ with $s/1000$ to produce the following transfer function:

$$H(s) = \frac{V_{\text{out}}}{V_{\text{in}}} = \frac{1}{\$^2 + 0.804\$ + 0.823}\bigg|_{\$=s/1000} = \frac{10^6}{s^2 + 804s + 823.10^3} \qquad (4.3)$$

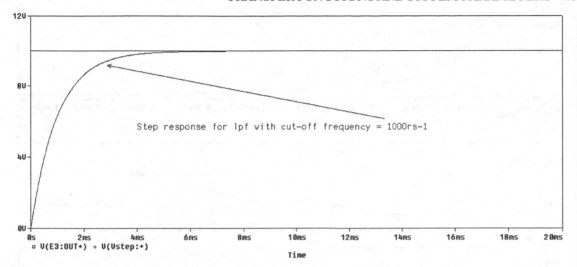

FIGURE 4.12: First-order step response

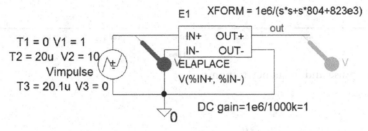

FIGURE 4.13: Impulse parameters

The maximum response occurs at the resonant frequency.

$$\omega_0^2 = 823 \text{ krs}^{-1} \Rightarrow f_0 = (\sqrt{\omega_0^2})/2\pi = \sqrt{823} \text{ kr}^{-1}/2\pi = 145 \text{ Hz}$$

However, this is not the same as the passband edge frequency. The gain at DC is 1 Meg/0.823 Meg = 1.21. Fig. 4.13 shows an **ELAPLACE** part with the transfer function entered in the **XFORM** as 1e6/($s*s + 804*s + 823K$). Select the **VPWL** part and enter the impulse input signal parameters to produce a 10 V pulse with a width of 2 us.

Set Analysis to **Analysis type: Time Domain (Transient), Run to time** = 100 ms, and **Maximum step size** = 1 us. Press **F11** to simulate. Separate the overlapping displays by creating two more extra plots using the **Alt PP** buttons twice. Delete and select the output variable you wish to move by selecting the **V(out)** variable and pressing the **Ctrl X** buttons. Paste the copied variable into the new plot with **Ctrl V**.

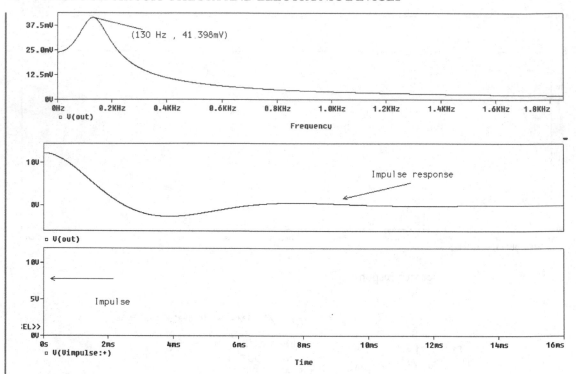

FIGURE 4.14: Impulse and frequency response

4.5.1 Unsynchronizing Probe Plots

On the top plot, select **Plot/Unsyncronize X-Axis** and press the **FFT** icon. Unsynchronizing the x-axis enables us to plot the frequency response but keep the other plots still in the time domain as shown in Fig. 4.14. You might need to increase the transient **Run to time** to 1 s in the analysis setup to achieve good frequency resolution.

Measure the frequency at the max of the display and the time between peaks in the middle plot (verify it is equal to $1/f_0$). Measure the overshoot, settling time, and natural frequency. There is an error at the beginning of the frequency response but see Section 4.7 for an explanation.

4.6 FIRST-ORDER LOW-PASS FILTER STEP AND IMPULSE RESPONSES

Impulse testing is useful for assessing system stability. An impulse is created using the **VPULSE** part by setting the period much larger compared to the pulse width as shown in Fig. 4.15, where the impulse is 10 ns wide and the period is 10 s.

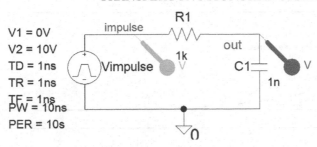

V1 = 0V
V2 = 10V
TD = 1ns
TR = 1ns
TF = 1ns
PW = 10ns
PER = 10s

FIGURE 4.15: LPF

PW	10n
PER	10
TD	1n
TF	1n
TR	1n
V1	0v
V2	10

FIGURE 4.16: Vpulse generator parameters

Simulation Settings - TRAN

General | Analysis | Configuration Files | Options | Data Collection | Probe Window

Analysis type:
Time Domain (Transient) ▼

Run to time: 100u   seconds (TSTOP)

Start saving data after: 0   seconds

Options:
☑ General Settings
☐ Monte Carlo/Worst Case
☐ Parametric Sweep
☐ Temperature (Sweep)
☐ Save Bias Point
☐ Load Bias Point

Transient options
Maximum step size: 100ns   seconds
☐ Skip the initial transient bias point calculation (SKIPBP)

Output File Options..

FIGURE 4.17: Transient analysis parameters

Enter the **VPULSE** generator parameters shown in Fig. 4.16 and set the **Transient Analysis** as shown in Fig. 4.17.

Transient Analysis parameters are shown in Fig. 4.17. After simulation, create separate display using **Alt PP** and place the output signal on top by applying **Ctrl X** on the selected output variable at the bottom left of the **Probe** output. The x-axis range is changed from a **Run to time** of 15 ms to 10 us so that the impulse signal and response can be observed clearly, as shown in Fig. 4.18.

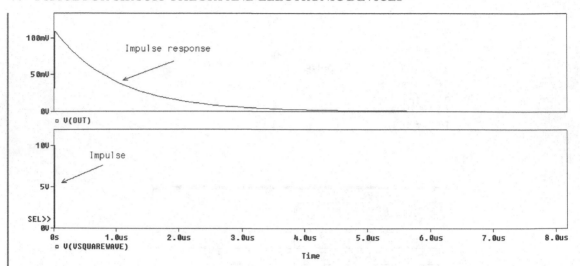

FIGURE 4.18: The impulse response

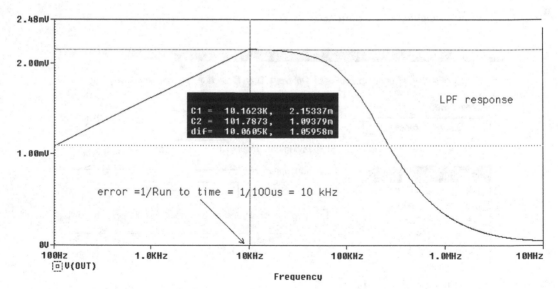

FIGURE 4.19: Frequency response

4.7   OBTAINING THE FREQUENCY RESPONSE FROM THE IMPULSE RESPONSE

The **F**ast **F**ourier **T**ransform **(FFT)** icon displays the frequencies contained in a complex signal but is also useful for obtaining the frequency response from the impulse response. The amplitude frequency response in Fig. 4.19 was obtained from the impulse response by selecting the **FFT** icon from the **Probe** screen toolbar. In this experiment, we wish to obtain the frequency response

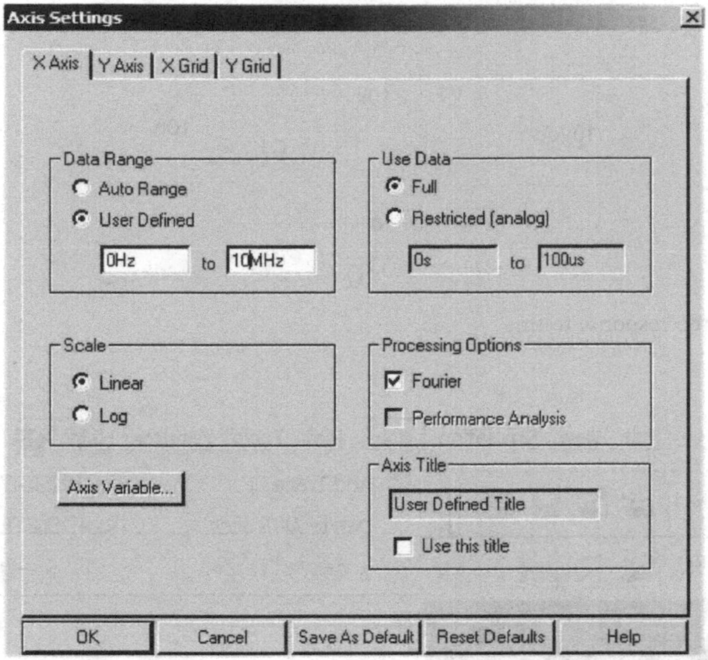

FIGURE 4.20: Changing the x-axis

for the low-pass filter by applying an impulse to the circuit. However, a small error will occur at the beginning of the frequency response. **Run to time** = 15 ms results in the incorrect display from 0 Hz to 66.66 Hz (66 Hz = 1/15 ms).

Making the **<u>R</u>un to time** larger increases the resolution in the frequency domain and also reduces the error at the start. Change the frequency axis by selecting an x-axis number and setting the parameters as shown in Fig. 4.20.

We should remember this technique when we cannot carry out a normal ac frequency response, for example, when a circuit contains switch components, such as in a switched capacitor circuit, and requires a transient analysis [ref: 1].

4.8   THE LOW-PASS *CR* FILTER STEP RESPONSE

From the step response, we may determine system parameters, such as rise time, and steady-state value. In second-order systems, we may measure overshoot, settling time, and natural frequency. Draw the schematic shown in Fig. 4.21 using a VDC part to provide a step input signal.

From **Analysis/Transient Analysis** menu, tick **S<u>k</u>ip initial transient solution** and set the **Run to time** to 1 ms. If we use a **VPULSE** generator instead of the VDC part, then make

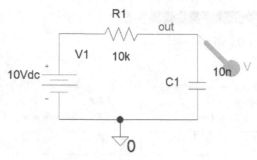

FIGURE 4.21: Step response testing

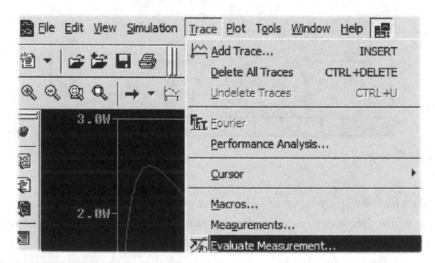

FIGURE 4.22: Evaluate measurement functions

the rise time, **TR**, and pulse width, **PW**, parameters wide enough to allow the system transients to die out.

4.9   RISE TIME

Leading-edge degradation in digital signals may cause problems so some means of assessing the degradation are required. Rise time is one method for assessing changes in the pulse leading edge degraded due to bad component layout, circuit design, or a channel with limited bandwidth. **Rise time** is the *time for the output signal to rise from 10% of the final value to 90% of the final value* and is expressed as $t_r = 2.2\tau = 0.35/f_c$. The **Evaluate measurement** function in Probe can be used to measure the rise time without using the cursors. Press **F11**, and select **Probe/Trace/Evaluate Measurement** menu as in Fig. 4.22.

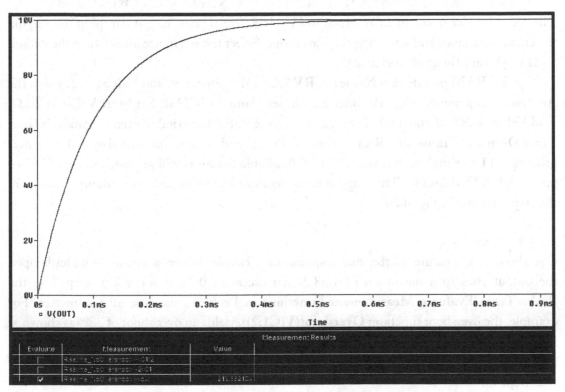

FIGURE 4.23: Evaluate measurement functions

FIGURE 4.24: Rise-time measurement using Evaluate Measurement function

Select **Risetime (1)** from the list of functions in Fig. 4.23 to place it in the **Trace Expression** box.

Replace the default "**1**" value in the function brackets with the capacitor voltage **V(out)** function, selected from the list on the left. Fig. 4.24 shows the **Evaluate Measurement** function evaluated at the bottom as 10% final value to 90% final value equal to $t_r = 2.2\tau = 2.2\,CR = 2.2*1e4*10e - 9 = 219$ us.

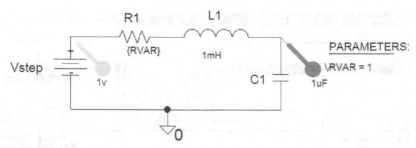

FIGURE 4.25: Second-order LCR circuit

4.10 STEP RESPONSE OF A SERIES-TUNED LCR CIRCUIT

The second-order resonant *LCR* circuit in Fig. 4.25 investigates "circuit damping" by varying the circuit resistance and observing the waveforms. Select the resistance and change the default 1 kΩ to {**Rvar**} (brackets included).

A **PARAM** part defines **Name1 = RVAR** for the resistance, and **Value1 = 1**. From the **Analysis Setup** menu, select **Parametric,** and set **Name** to RVAR. Set **Start Value = 20 Ω**, **End Value = 200 Ω** and a 60 Ω resistance increment. Set the **Analysis** tab to **Analysis type: Time Domain (Transient), Run to time = 15** ms, and **Maximum step size = 1** us. After pressing **F11** to simulate, a small screen of **Available Sections** will appear, so select **OK,** or press the **ENTER** button. The range of damping from underdamped to overdamp is shown in the step response in Fig. 4.26.

4.10.1 Overshoot

Overshoot is a measure of the step response of a circuit. When a circuit is underdamped the output rises to a maximum of 1.38 V, an excess of 0.38 V for a 1 V step. Use the **Probe/Trace/Evaluate Measurement Functions** in **Probe** to measure all parameters. For example, the overshoot function: **Overshoot(V(C1:2)),** yields an over shoot of 38% as shown at the bottom of Fig. 4.26. You should also notice at the bottom of the display symbols representing each trace. Selecting any one of these symbols shows the information for a particular value of the swept variable.

4.11 EXERCISES

(1) Use an **ABM E** part to create the integrator in Fig. 4.27. The input data is applied using a **VPWL** part, or a **VPULSE** part with ±1 V amplitude, **PW =**1 ns, and **PER = 2** n. Here, $\tau = CR = 1$ kΩ × 2 pF = 2 ns.

(2) An oscilloscope probe reduces the effect of the input oscilloscope capacitance on a signal. Without such a probe, a step or pulse signal has the rise time degraded. The 10:1 oscilloscope probe in Fig. 4.28 is calibrated by attaching the probe end onto the

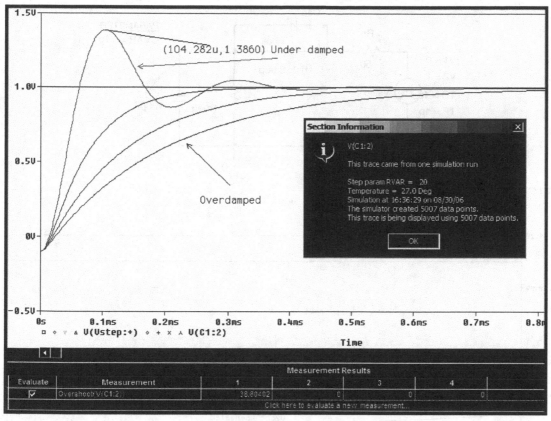

FIGURE 4.26: Range of damping coefficients

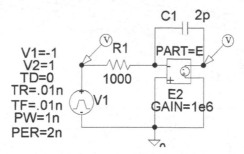

FIGURE 4.27: Integration using the E part

reference square wave signal on the front panel of the oscilloscope. In normal usage, the probe tip variable capacitance is adjusted by inserting a small plastic trimming tool and adjusting it to produce the best displayed square wave. The best display is where the applied signal looks like the original reference but reduced in amplitude by a factor of 10.

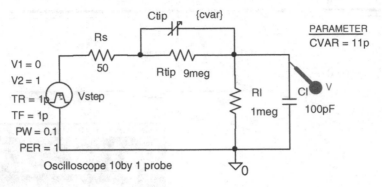

FIGURE 4.28: Oscilloscope probe

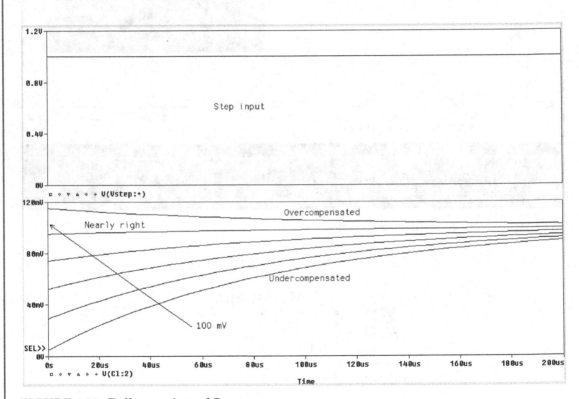

FIGURE 4.29: Different values of Cvar

This is simulated in PSpice using a **PARAM** part to vary Ctip from 1 pF to 30 pF in steps of 1 pF. Use the **Evaluate Measurement** function **Risetime (V (Cl:1))** from the list. Determine from Fig. 4.29, the best value for **cvar**. The best value occurs when the displayed scope signal is a square wave (This occurs when the output voltage is one tenth of the input voltage—hence the probe name). A useful exercise is to plot the

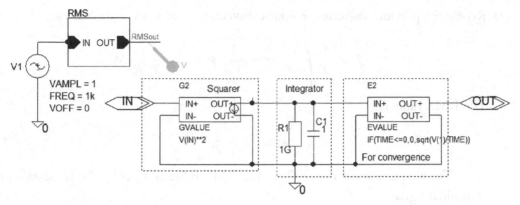

FIGURE 4.30: RMS meter in block form

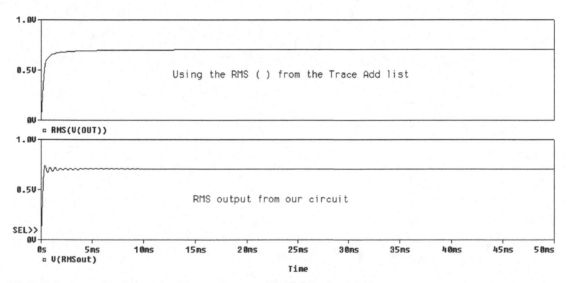

FIGURE 4.31: Comparison of Probe function RMS () to the RMS meter

frequency response for this Probe circuit using a **VAC** part and run an AC analysis for the different **cvar** capacitor values. What should you observe (high-pass or low-pass response?)

(3) The true RMS meter in Fig. 4.30 is constructed from **ABM** parts (**GVALUE** and **EVALUE**). Apply a 1 kHz, 1 V sinusoidal signal using a **VSIN** part. To square the input voltage, enter **EXP = V(in)*V(in), or (V(in)**2)** into the **GVALUE** part. The integrator is followed by an **EVALUE** part using the **IF then ELSE** statement as: **IF (TIME<=0, 0, sqrt(V(1)/TIME))**. This avoids convergence problems, but also gets the square root of the voltage on the wire labeled 1. Run a transient analysis for 50 ms. Fig. 4.31 compares the RMS output to the probe function RMS (**V(RMSout)**). Both eventually settle down to the same output voltage of 0.707 V.

(4) Repeat the previous exercise for a pulse waveform. The power of a pulse waveform is

$$P = \frac{1}{T} \int_0^\tau \frac{V^2}{R} dt = \frac{1}{TR}(V^2 t)_0^\tau = \frac{\tau}{T}\frac{V^2}{R} \text{ W} \qquad (4.4)$$

The RMS value is

$$V_{\text{RMS}} = \sqrt{\frac{\tau}{T}} V \text{ V} \qquad (4.5)$$

Note: For a square wave $\tau = T/2$, so the RMS value is $V/\sqrt{2}$—the same as a pure sinusoidal signal.

CHAPTER 5

AC Circuits and Circuit Theorems

5.1 AC CIRCUIT THEORY

Capacitors and inductors have frequency-dependant reactive properties making them useful for modifying the frequency content of signals. In this chapter, we apply circuit theorems to a range of circuits such as series and parallel resonant circuits used in many communication circuits where important concepts, such as selectivity and Q-factor are examined.

5.2 CAPACITORS

Two metal plates of area of A and separated by an insulator thickness d and absolute permittivity ε has capacitance:

$$C = \frac{\varepsilon A}{d} \qquad (5.1)$$

Connecting a voltage V across a capacitor deposits a charge, Q coulombs, on its plates (Charles Coulomb 1738–1806). Since current is proportional to the rate of change of charge with time and $Q = CV$, therefore

$$i = \frac{dQ}{dt} = C\frac{dV}{dt} \qquad (5.2)$$

A voltage $V_c(t) = V_m \sin \omega t$ across a capacitor C produces current that leads the voltage by $90°$.

$$i = C\frac{dV_c}{dt} = C\frac{d(V_m \sin \omega t)}{dt} = \omega C V_m \cos \omega t = \frac{V_m}{1/\omega C} \cos \omega t = \frac{V_m}{X_C} \cos \omega t = I_m \cos \omega t$$

$$(5.3)$$

Rearranging (5.3) yields the capacitive reactance:

$$X_C = V_m/I_m = 1/\omega C \ \Omega \qquad (5.4)$$

Capacitive reactance decreases with frequency increase and is an open circuit at DC. We may express the current–voltage relationship as susceptance and is the reciprocal of reactance defined as $B_c = I_m/V_m = \omega C$ Siemens.

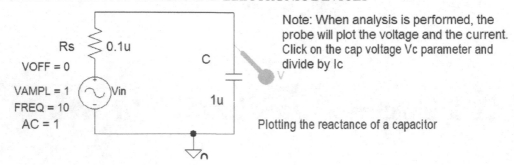

FIGURE 5.1: Capacitive reactance

5.2.1 Capacitive Reactance Plot

Capacitive reactance is investigated using the schematic in Fig. 5.1, where a **VAC** generator part is connected to the capacitor via a small resistance that represents the source resistance of the voltage source.

Set the Analysis tab to **Analysis type: AC Sweep/Noise, AC Sweep Type** to **Linear, Start Frequency** = 0.01, **End Frequency** = 10, **Points/Decade** = 10,000 (this large value gives a smoother reactance plot). There are two techniques for plotting signals in Probe. The first technique is to place markers on the schematic, which, after simulation, automatically plots the required voltage or current waveforms. The second technique uses no markers but the variables are selected after pressing the **Trace/ Add Trace** menu from the Probe screen shown in Fig. 5.2.

Alternatively, press the **Insert** button and in **Trace Expression** box, select the variables **V(C: 2)/I(C)** from the list to plot positive capacitive reactance. When the plot shown in Fig. 5.4, select an *x*-axis parameter or the space beside it, and it will open the menu shown in Fig. 5.3.

Change the Scale to **Log** and add a title by selecting **Use this title** and typing a name in the **Axis Title** box.

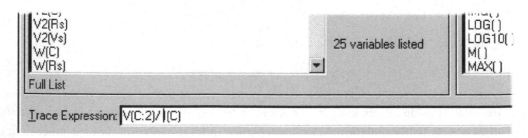

FIGURE 5.2: Select the voltage variable

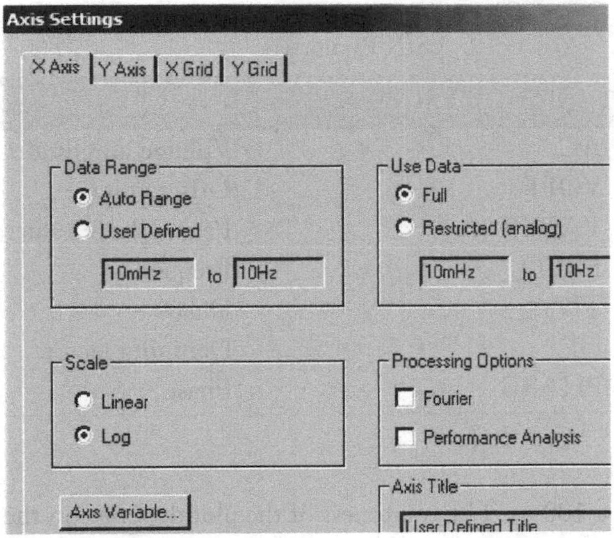

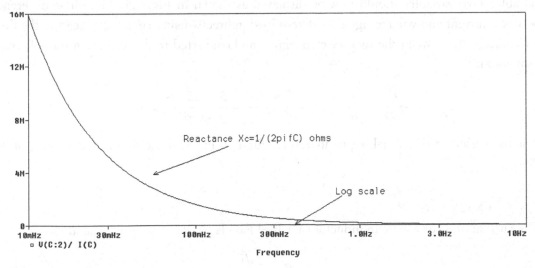

FIGURE 5.3: Changing the *x*-axis to logarithmic

5.2.2   Capacitor Current and Voltage Waveforms

To observe the phase relationships between the capacitor current and voltage, we need to carry out a transient analysis and measure the time between the peak values. The **VAC** generator is replaced by a **VSIN** generator whose parameters are shown in Table 5.1.

Enter the **VSIN** generator parameters: **VOFF = 0**, **VAMPL = 1** and **FREQ = 10**. **Set the Analysis** tab to **Analysis type: Time Domain (Transient)**, **Run to time = 0.2**, and

FIGURE 5.4: Capacitive reactance

TABLE 5.1: VSIN Parameters

PARAMETER	MEANING
AC	Voltage amplitude
VOFF	Offset voltage
VAMPL	Peak voltage value
FREQ	Frequency
TD	Delay
DF	Damping factor
PHASE	Phase

Maximum step size = 100 us. The coarseness of the plot depends on the **Maximum step size** which should be decreased whenever you observe triangulation, or straight lines, where they should be curved. To plot microamps and volts on the same plot, requires adding a different y-axis for each signal, otherwise the current signal will appear to be zero. This is fixed by adding a new Y-axis from the **Plot** menu and selecting **Add Y Axis (or Ctrl Y)**. Select the left-hand side of the inner vertical Y-axis and you should see the **Y-axis** number is 2.

Double chevrons $\gg$ on the y-axis indicates which axis is selected. From the **Trace** menu, select **Add,** or press the icon ⬚ and insert the capacitor voltage **V(C:2)** in the **Trace Expression** box. **Lclick** on the left-hand side of the outer vertical y-axis (The added axis), and observe the **Y-axis** number is 1. Press the insert button (or select the insert icon) and select the **I(C1)** variable. Two variables should now be displayed as shown in Fig. 5.5. The phase difference between current and voltage signals is determined indirectly using cursors to measure the time difference, Δt, between the two selected signals and converted to degrees using the following expression:

$$\frac{T}{360} = \frac{\Delta t}{\theta} \Rightarrow \theta = \frac{360.\Delta t}{T} = \frac{360.25 \text{ ms}}{100 \text{ ms}} = 90° \tag{5.5}$$

T is the period of the signal (equal to 100 ms for a 10 Hz a sinusoid) and has a phase of 2π (360°).

5.3 INDUCTORS

The current and voltage in an inductor (Michael Faraday 1791–1867) are related:

$$E = L\frac{di}{dt} \tag{5.6}$$

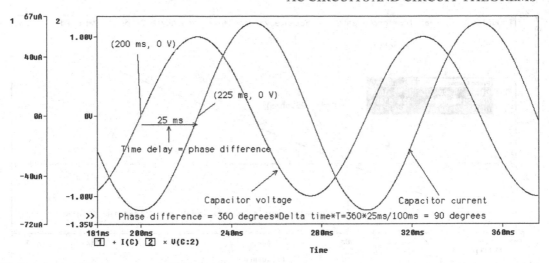

FIGURE 5.5: Capacitor $V-I$ Phase relationships

A voltage is produced *only* when the current is changing. We obtain an expression for inductive reactance by considering the inductor voltage for a sinusoidal current $i(t) = I_m \sin 2\pi f t$ as

$$E = L\frac{dI_m \sin 2\pi f t}{dt} = \omega L I_m \cos 2\pi f = I_m X_L \cos \omega t \Rightarrow X_L = \frac{E}{I_m} = \omega L \qquad (5.7)$$

The voltage leads the current by 90° and the reciprocal of reactance is susceptance $B_L = 1/\omega L$ Siemens. Fig. 5.6 shows a small resistance added since PSpice inductors are ideal and possess no winding resistance. Without this resistance you will get an error message: **error Voltage source and/or inductor loop involving V_Vs.** Set the **Analysis** tab to **Analysis type: AC Sweep/Noise, AC Sweep Type** to **Linear, Start Frequency** = 0.1, **End Frequency** = 4, **Points/Decade** = 1000.

Delete the current marker and press **F11 to simulate.** The Probe screen should now show a straight horizontal line. **Lclick** the plotted variable **V(L1:1)** and in the **Trace Expression** box,

Note: When ac analysis is performed, probe will plot the voltage VL and the current IL. Click on the inductor voltage VL and divide by IL Plotting the reactance of an inductance

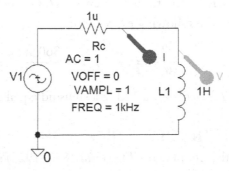

FIGURE 5.6: Schematic for plotting inductive reactance

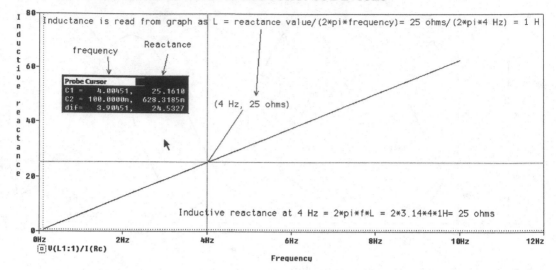

FIGURE 5.7: Inductive reactance plot

divide the variable by the inductor current i.e. **V(L1:1)/I(L1)**. Change x-axis **Scale** to **Linear** scale by selecting the **log/lin** icon. This makes the inductive reactance ωL plotted as a linear increase with frequency shown in Fig. 5.7. The impedance of an ideal inductor (Joseph Henry 1797–1878) is zero at zero frequency (DC) and is an open circuit at infinity frequency.

5.3.1   Inductor Signal Phase Measurement
Place a current marker on the output. Set the **Analysis** tab to **Analysis type: Time Domain (Transient), Run to time** = 10 ms, and **Maximum step size** = 1 us. Press **F11** to simulate and display the inductor current and voltage, as shown in Fig. 5.8. However, it is necessary to use the previous procedure for measuring capacitor V–I phase relationships i.e. in Probe, add a new axis (Ctrl Y) in order to display the current with an appropriate range.

The phase difference between current and voltage signals is determined indirectly using cursors to measure the time difference, Δt, between the two selected signals and converted to degrees using the following expression:

$$\frac{T}{360} = \frac{\Delta t}{\theta} \Rightarrow \theta = \frac{360 \times \Delta t}{T} = \frac{360 \times 250u}{1000us} = 90° \qquad (5.8)$$

The period T is 1 ms for a 1 kHz a sinusoid equal to a phase of $2\pi \, 360°$.

5.4   AC CIRCUIT THEOREMS
Thévenin's theorem (Leon Thévenin 1857–1926) states that any two-terminal circuit may be replaced by an equivalent circuit comprising an impedance measured across the open-circuit

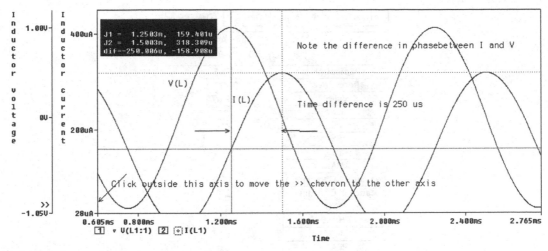

FIGURE 5.8: Inductor $V-I$ phase plot

output determinals, but with all sources replaced by their internal impedances, and placed in series with the voltage measured across the output terminals when open circuited.

5.4.1 Thévenin's Theorem

The large value resistance ZL in Fig. 5.9 represents an open-circuit load in order to measure values for the Thévenin equivalent circuit. Apply a 10 V **VAC** part and measure the current and voltage in each loop using the **IPRINT and VPRINT** parts (a printer symbol). Loop currents are logged by ammeters, mesh1, and mesh2 and the measurements from these virtual instruments are contained in an output file accessed from the **PSpice/View Output File** menu (see Fig. 5.13).

DLclick the AC voltage source $V1$, and enter 10 V and phase angle $= 0$ in the **Edit Properties** spreadsheet. **Vprint1** (A printer symbol in the **special.olb** library) is used to record voltage values in the **Output** file. Select the meter and enter the **Vprint1** voltmeter parameters

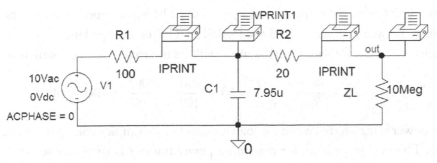

FIGURE 5.9: Two-loop circuit

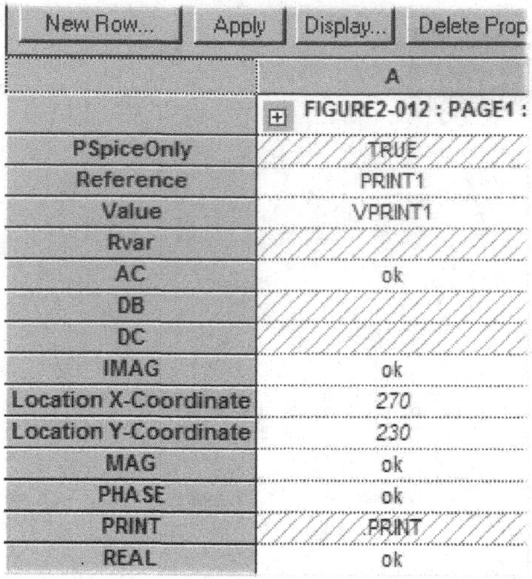

	A
	⊞ FIGURE2-012 : PAGE1 :
PSpiceOnly	TRUE
Reference	PRINT1
Value	VPRINT1
Rvar	
AC	ok
DB	
DC	
IMAG	ok
Location X-Coordinate	270
Location Y-Coordinate	230
MAG	ok
PHASE	ok
PRINT	PRINT
REAL	ok

FIGURE 5.10: The VPRINT1 parameters

as shown in the spreadsheet shown in Fig. 5.10. Type **ok,** or **y**, beside a parameter you wish displayed in the output file.

The **IPRINT1** part measures the current in each loop, and is placed *in series* with a component. Select the part and fill in the required parameters as per the **VPRINT** part. Specify the ammeter parameters by typing ok beside each parameter required in the output file. Set the **Analysis** tab to **Analysis type: AC Sweep/Noise, AC Sweep Type to Linear, Start Frequency = 100, End Frequency = 100, Points/Decade = 1**. The analysis is carried out when the start and end frequencies are equal, i.e. at one frequency only. In this example, the capacitance $C1$ has a reactance of 200 Ω at a frequency of 100 Hz.

5.4.2 Thévenin Impedance

Apply Thévenin's theorem to the left of the load ZL, which is a large resistance (10 Meg) representing an open circuit. Create the equivalent circuit and place it on the same schematic area. Replacing ZL with the complex conjugate of the Thévenin impedance ensures maximum power transfer between the circuit and the load. The Thévenin impedance is R_2 in series with R_1 in parallel with X_{c1}. Replace the V_1 supply with its internal resistance, which is zero.

$$Z_{\text{TH}} = R_2 + \frac{R_1(-jX_{C1})}{R_1 - jX_{C1}} = 20 + \frac{100(-j200)}{100 - j200} = 100 - j40 \; \Omega \qquad (5.9)$$

Maximum power is transferred when the load is equal to the complex conjugate of the Thévenin impedance. The load impedance for maximum power transfer is therefore set to the complex conjugate of the Thévenin impedance i.e. $Z_{\text{LOAD}} = 100 + j40 \; \Omega$. The inductance is calculated

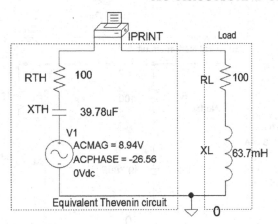

FIGURE 5.11: Thévenin equivalent circuit

from the reactance as

$$X_L = 2\pi f L = 40\ \Omega \Rightarrow L = \frac{40}{2\pi\,100} = 0.0637\ \text{H} \tag{5.10}$$

PSpice only accepts capacitance in farads and inductance in henries, so we must calculate, from the capacitive and inductive reactance, the equivalent value in farads, or henries, at the frequency of simulation. The Thévenin impedance is formed from a 100 Ω resistance in series with a capacitance whose reactance is $-j40\ \Omega$. We determine the capacitance from the capacitive reactance at 100 Hz as

$$X_C = \frac{1}{2\pi f C} = 40\ \Omega \Rightarrow C = \frac{1}{2\pi f X_C} = \frac{1}{2\pi\,100.40} = 39.78\ \text{uF} \tag{5.11}$$

Note: Thévenin circuit components in polar form are converted to rectangular form. The resistive part (the real part) is $z\cos\theta = 100\ \Omega$, and the reactance (the imaginary part) is $z\sin\theta = 40\ \Omega$.

5.4.3   Thévenin Voltage
The Thévenin voltage is calculated by applying the potential divider principle as

$$V_{\text{TH}} = \frac{-jX_{C1}}{R1 - jX_{C1}} V_1 = \frac{-j200}{100 - j200}10 = 8.94\angle -26.56\ V \tag{5.12}$$

Place the Thévenin equivalent circuit shown in Fig. 5.11 beside the original schematic.

5.5   NORTON EQUIVALENT CIRCUIT
The Norton impedance (Edward Norton 1898–1983) is the same as the Thévenin impedance, i.e. $Z_N = Z_{\text{TH}}$. The Norton current (the short-circuit current) is obtained from the Thévenin

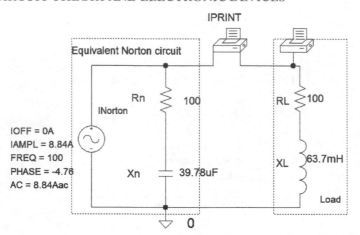

FIGURE 5.12: Norton equivalent circuit

equivalent circuit as

$$I_N = \frac{E_{TH}}{Z_{TH}} = \frac{8.84\angle - 26.56}{100\angle - 21.8 \; \Omega} = 8.84\angle - 4.76 \; \text{mA} \qquad (5.13)$$

Alternatively, replace the load in Fig. 5.12 with a very small 1 uΩ resistance and measure the short-circuit current. The Norton equivalent short-circuit current generator is an **ISIN** part with parameters as shown in Fig. 5.12. You may draw the Norton equivalent circuit beside the Thévenin equivalent.

Select **Simulate** from the **Analysis** menu, or press **F11**.

5.5.1   The Output File
Examine the output file shown in Fig. 5.13 selected from the **PSpice/View Output File** menu (or from **Probe/View**). The required ac information is at the end of the file. **FREQ** is a frequency of 1.000E+02 (100 Hz), and **IM** (V_ammeter1) reads 1.989 V.

5.6   AC MESH AND NODAL ANALYSIS
Apply mesh and nodal analysis to the main circuit and calculate the loop currents and nodal voltages. Remember to replace ZL by the complex conjugate of the Thévenin impedance before any numerical calculations. From nodal analysis, calculate the voltage at the junction of R1 and R2 and compare to the value in the output file.

5.7   EXERCISES
(1) In Fig. 5.14. Measure all loop currents and node voltages and compare to values calculated using mesh or nodal analysis.

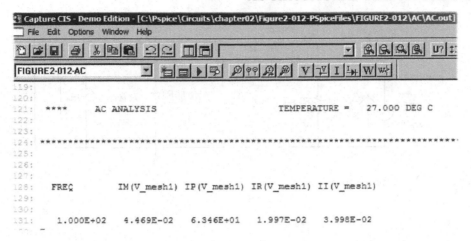

FIGURE 5.13: The output file

(2) Investigate the schematic in Fig. 5.15 and measure all loop currents and node voltages.

(3) Develop a Norton equivalent circuit with respect to the load Z_L, for the circuit shown in Fig. 5.16. Hence, calculate component for the equivalent circuit and a value for Z_L, for maximum power transfer. If the load ZL is replaced by a voltage source V_2, apply nodal analysis and obtain an equation in matrix form relating the circuit components and parameters. $R_1 = 100 \, \Omega$, $R_2 = 20 \, \Omega$, $X_L = j618 \, \Omega$, $V_1 = 10 \sin 2\pi 100t$ V.

(4) For the network in Fig. 5.17, write expressions for the currents in X_{L1} and X_{L2} in a matrix form. Calculate the node voltage at the junction of X_{L1} and X_{L2} using nodal analysis. $R_1 = 100 \, \Omega$, $X_{L1} = X_{L2} = j100 \, \Omega$, $V_1 = 1 \angle 0$ V, $V_2 = 2 \angle 0$ V.

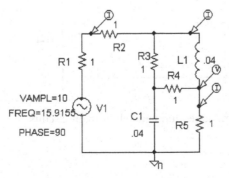

FIGURE 5.14: Mesh problem

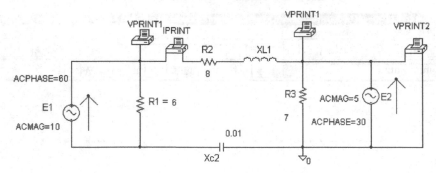

FIGURE 5.15: Mesh problem

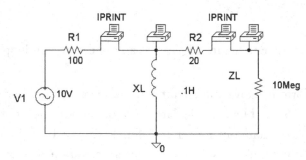

FIGURE 5.16: Mesh problem

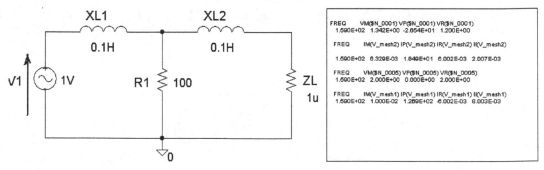

FIGURE 5.17: Mesh problem: Series and parallel resonant circuits

CHAPTER 6

Series and Parallel-tuned Resonance

6.1 RESONANCE

The resonance phenomenon occurs in a second-order system containing capacitors and inductors. A circuit containing resistance, capacitance, and inductance will become purely resistive at the resonant frequency when the inductive reactance equals the equal capacitive reactance. **Selectivity** is the ability of the resonant circuit to extract the resonant frequency and attenuate other frequencies. Selectivity is measured by a Q-factor, which, for a series-tuned circuit, is the ratio of the inductive (or capacitive reactance) to the total resistance in the circuit (this includes source and load resistances). It is a measure of how well the circuit extracts a band of frequencies with little attenuation but rejects other frequencies outside this band. Selectivity is also dependent on the inductor-capacitor ratio.

A **high Q-factor value** means **high selectivity** and a low Q-factor means low selectivity. For good **series-tuned** selectivity, the circuit must be fed from a **voltage source** because the source resistance feeding the circuit has minimum resistance and hence has minimum effect on the Q-factor (Ideal voltage source impedance is zero). Any series resistance added will reduce the overall selectivity. A *parallel-tuned circuit* is fed from a *current source* because it has a high source resistance and hence minimum loading. Any external resistive loading reduces the overall selectivity of the parallel-tuned circuit.

6.2 SERIES-TUNED CIRCUIT

Fig. 6.1 shows a series-tuned circuit with inductance, capacitance, and resistance R representing source and coil resistances. The total circuit impedance is

$$Z = E_S/I_S = R + j(X_L - X_C) \qquad (6.1)$$

The total reactance is zero at the resonant frequency f_0 (j terms are zero).

$$(X_L - X_C) = 0 \Rightarrow X_L = X_C \text{ or } \omega L = 1/\omega C \qquad (6.2)$$

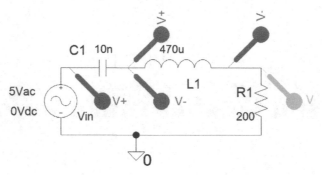

FIGURE 6.1: A series-tuned circuit

Cross-multiplying (6.2) yields an expression for resonance as

$$\omega^2 = \frac{1}{LC} \Rightarrow \omega = \frac{1}{\sqrt{LC}} \Rightarrow f_0 = \frac{1}{2\pi\sqrt{LC}} \qquad (6.3)$$

Below resonance, the circuit impedance is capacitive i.e. a leading phase angle. Above resonance the circuit is inductive i.e. lagging phase angle and at f_0, the phase is zero (j term is zero).

6.2.1 Current Response

Maximum current flows at resonance when the circuit is purely resistive. We can measure the -3 dB bandwidth from the current response in Fig. 6.2 by expanding the area around the -3 dB area using the magnifying glass icon. However, it may be necessary to simulate again but with a smaller frequency range for better resolution.

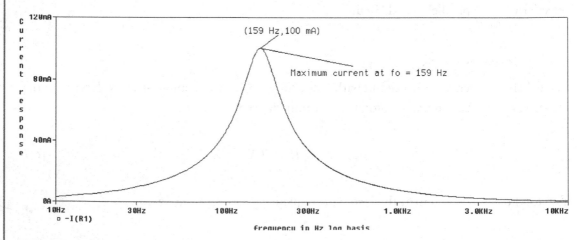

FIGURE 6.2: Current response

6.3   EXAMPLE

A **VSIN** generator $E_S = 10 \sin 2\pi f t$ V is connected to a series-tuned circuit. $L = 470$ uH and $C = 10$ nF. If the total resistance is 4 Ω, calculate and verify by simulation: The frequency of resonance f_0, the phase of the circuit a decade above and below f_0, and the value of the current at f_0.

6.3.1   Solution

Resonance occurs at $f_0 = 1/2\pi \sqrt{LC} = 73 \cdot 45$. The phase is $\theta = \tan^{-1}(X_L - X_C)/R = \tan^{-1}(2\pi f L - 1/2\pi f C)/R$. The phase a decade above resonance $f = 734.5$ kHz is $\theta = \tan^{-1}(2167 \cdot 9 - 21 \cdot 679)/4 = 89°53'$. The phase a decade below resonance $f = 7.345$ kHz is $\theta = \tan^{-1}(21.679 - 02167.9)/4 = -89°53$. The impedance at resonance is

$$Z = R + j(X_L - X_C) = 4 + j(216 \cdot 79 - 216 \cdot 79) = 4 + j0 \ \Omega$$

The current at resonance is calculated by dividing the voltage by the impedance:

$$i_s(f_0) = E_S/R = 10 \sin 2\pi f t/4 = 2 \cdot 5 \sin 2\pi f t$$

6.4   Q-FACTOR

The Q-factor is a measure of SELECTIVITY which is the ability of a resonant circuit to select out a band of frequencies whilst attenuating other frequencies outside this band. The Q-factor is obtained from the ratio of the energy stored in the coil reactance to the energy dissipated in the circuit resistance. Typical coil Q-factors' range is 20–80, but capacitor Q-factors are much bigger with values in the thousands. Because of the large difference in Q-factor values, the resonant circuit selectivity is determined mainly by the inductor Q-factor.

$$Q = \frac{\text{Energy stored}}{\text{Energy dissipated per cycle}} = \frac{\omega L}{R} = \frac{\text{reactance}}{\text{resistance}} = \frac{1/\omega C}{R} = \frac{1}{\omega C R} \qquad (6.4)$$

6.4.1   The −3 dB Bandwidth

To compare the selectivity of resonant circuits, we introduce the −3 dB bandwidth defined as the difference in two frequencies, f_1, and f_2, frequencies where the output voltage falls by 3 dB from the maximum at f_0. At these frequencies, the resistance in the circuit is equal to the total reactance i.e. $R = X_T = X_L - X_C = 2\pi f_i L - 1/2\pi f_i C$. However, the total reactance decreases on either side of the resonant frequency, so we equate the resistance to the reactance on either side and solve. Capacitive reactance below resonance is greater than inductive reactance i.e. $R = 1/\omega_1 C - \omega_1 L = \omega_2 L - 1/\omega_2 C$ so multiplying by $\omega_1 C$ yields $1 - \omega_1^2 LC = \omega_1 \omega_2 LC - \omega_1 C/\omega_2 C$. At resonance, this reduces to $LC = 1/\omega_0^2$. To obtain an expression for the quality factor, replace LC to yield:

$$1 - \frac{\omega_1^2}{\omega_0^2} = \frac{\omega_1\omega_2}{\omega_0^2} - \frac{\omega_1}{\omega_2} = 1 + \frac{\omega_1}{\omega_2} = \frac{\omega_1\omega_2}{\omega_0^2} + \frac{\omega_1^2}{\omega_0^2} \qquad (6.5)$$

$$\frac{\omega_1 + \omega_2}{\omega_2} = \frac{\omega_1(\omega_1 + \omega_2)}{\omega_0^2} = \omega_0^2 = \omega_1\omega_2 \qquad (6.6)$$

$$f_0^2 = f_1 f_2 \Rightarrow f_0 = \sqrt{(f_1 f_2)} \qquad (6.7)$$

The relationship between f_0, BW, and Q is

$$Q = f_0 / BW \qquad (6.8)$$

Any resistance added to a series-tuned circuit **decreases the Q-factor**, so the circuit must be fed from a voltage source and not a current source. However, a real voltage source has resistance and will reduce the overall selectivity.

6.5   VOLTAGES ACROSS L AND C AT RESONANCE

We will now see how we can get voltage magnification in passive resonant circuits—a phenomenon called Q-magnification. The current at the resonant frequency, f_0, is

$$i = E_S / R \qquad (6.9)$$

Hence, the inductor voltage is

$$V_L = ij X_L \qquad (6.10)$$

Substituting (6.9) into (6.10) yields the magnitude of the inductor voltage as

$$|V_L| = E_S(X_L)/R \qquad (6.11)$$

Since the quality factor is $Q = X_L/R$, then

$$|V_L| = QE_S \qquad (6.12)$$

The voltage across the inductor is Q times the source voltage at resonance (Nikola Tesla 1856–1943). A similar analysis for the capacitor voltage $V_c = -ij X_C$ yields

$$|V_c| = E_S X_C / R = QE_S \qquad (6.13)$$

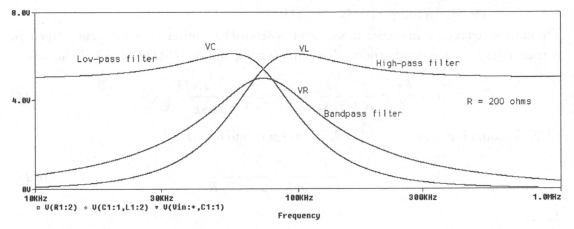

FIGURE 6.3: Voltage response across $R = 200\,\Omega$, C, and L

The capacitor voltage is Q times the applied voltage and occurs across L and C at resonance forming the basis of the Q-meter. For example, if $E_S = 10$ V and $Q = 50$, then the voltage V_L across the coil is 500 V. Q-magnification results in peak voltages as shown in Fig. 6.3 for $R = 20\,\Omega$, $C = 10$ nF, and $L = 470$ uH. The peaks will tend to merge at the resonant frequency for Q-factors greater than 10, but will separate for low value Q-factors.

The voltage across the inductance falls off at low frequencies displaying *high-pass filter* characteristics, whilst the voltage across the capacitor rolls-off at high frequencies giving a *low-pass filter* response. The resistor voltage, on the other hand, has a *bandpass* response. The voltages across C and L are picked up using **differential voltage markers** from the **Marker** menu. The peak separation is evident only for low Q-factor values (this is achieved by increasing the resistance to 200 Ω). *Note*: An alternative to plotting the resistor voltage is to including the node at either end of the resistor in the trace expression dialogue box in the Probe screen as $V(R1:2)\ -V(R1:1)$. These are the alias voltages at either end of the resistance.

To obtain an expression for the frequency at which the peak capacitor voltage occurs means differentiating the capacitor voltage with respect to frequency and equating the result to zero. Another useful facility in probe is the **Append** command found in the **File** menu where you may combine simulation data from another simulation e.g. **example.dat**. However, the appended file must have a different name, and have the same x-axis settings. Measure the peak voltages V_C and V_L. The maximum voltage is Q times the source voltage, so we may calculate a value by dividing the maximum voltage by the source voltage. To change the cursor from one trace to another, click the diamond/square symbol at the bottom of the Probe output besides the measured parameter.

6.6   UNIVERSAL RESPONSE CURVE

The normalized current in a series-tuned circuit is defined by a universal response curve equation that contains two factors: a fractional de-tuning factor x and the Q-factor and is defined as

$$i = \frac{E_S}{Z} = \frac{E_S}{R + j(X_L - X_C)} = \frac{E_S/R}{1 + j(X_L - X_C)/R} \tag{6.14}$$

At f_0, maximum current is $i_0 = E_s/R$, so substitute into (6.14)

$$i = \frac{i_0}{1 + j(\omega L - 1/\omega C)/R} \tag{6.15}$$

Take out ωL

$$i = \frac{i_0}{1 + j\frac{\omega L}{R}(1 - \frac{1}{\omega^2 LC})} = \frac{i_0}{1 + j(\frac{\omega_0}{\omega})\frac{\omega L}{R}(1 - \frac{1}{\omega^2 LC})} = \frac{i_0}{1 + j(\frac{\omega_0 L}{R})\frac{\omega}{\omega_0}(1 - \frac{1}{\omega^2 LC})} \tag{6.16}$$

where $\omega_0^2 = 1/LC$ and $Q_0 = \omega_0 L/R$.

$$i = \frac{i_0}{1 + j Q_0 (\omega/\omega_0 - \omega_0/\omega)} \tag{6.17}$$

Normalize the current by dividing by i_0 to give the magnitude and angle as

$$\left| \frac{i}{i_0} \right| = \frac{1}{\sqrt{1 + Q_0^2 (\omega/\omega_0 - \omega_0/\omega)^2}} \angle -\tan^{-1} Q_0 (\omega/\omega_0 - \omega_0/\omega) \tag{6.18}$$

6.7   SELECTIVITY OF A SERIES-TUNED RESONANT CIRCUIT

We use a **Param** part from the **special.olb** library to investigate how resistance changes the circuit bandwidth as shown in Fig. 6.4. Select the **PARAM** part, **Rclick,** and select **Edit**

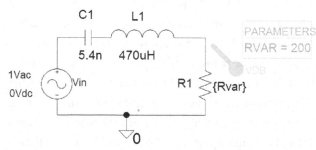

FIGURE 6.4: Selectivity testing

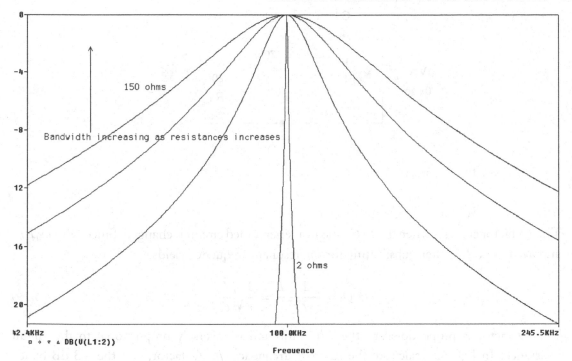

FIGURE 6.5: Selectivity versus circuit resistance

Properties. Enter a new row in the spreadsheet and place Rvar in the first column and 1 in the second column.

   DLclick the $R1$ value and enter {Rvar} (Make sure to include the brackets {}). From the **Analysis Setup** menu, click **Parametric,** and set the following parameters: **Global Parameter, Sweep Type = Linear, Name** = Rvar, **Start Value** = 1, **End Value** = 200, **Increment** = 50. Enter **Rvar (no curly brackets)** in the name box, and set the **Start, End,** and **Increment** to: 50, 200, and 50. Click on **AC Sweep** and set the **Analysis** tab to **Analysis type: AC Sweep/Noise, AC Sweep Type to Linear, Start Frequency** = 50k, **End Frequency** = 150k, **Points/Decade = 1000** to produce the response shown in Fig. 6.5.

   Measure the -3 dB bandwidth for the biggest resistance and calculate the Q-factor value. Compare to the measured value. The resistance determines the selectivity because the larger the resistance, the larger the bandwidth but a smaller Q-factor results. The bandwidth is 1 kHz for a resonant frequency $f_0 = 100$ kHz, and the loaded Q-factor is 100 (remember BW = f_0/Q_L). Set the **Start Frequency = 10k, End Frequency = 1000k** and **Points/Decade = 1000**. For detailed information around the resonant frequency region, choose a **Linear** sweep in the set-up menu and reduce the frequency range between 90 kHz and 110 kHz and simulate. Reducing the frequency axis in Probe will not increase the resolution.

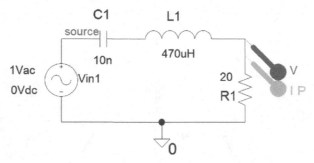

FIGURE 6.6: Phase marker

6.7.1 L/C Ratio and Selectivity

The Q-factor changes when the L/C ratio in a series-tuned circuit is changed. Since $Q = \omega_0 L/r$ and $\omega_0 = 1/\sqrt{LC}$, then substituting for the resonant frequency yields

$$Q = \frac{1}{\sqrt{LC}}\frac{L}{r} = \frac{1}{r}\sqrt{\frac{L}{C}} \tag{6.19}$$

The Q-factor is proportional to the L/C ratio and is inversely proportional to the circuit resistance. In Fig. 6.6, calculate the resonant frequency f_0, Q-factor, and the -3 dB bandwidth for $R = 2\Omega$ and $200\ \Omega$, and verify these values from the current amplitude (I_R) and phase responses (I_p) for $R = 2\Omega$. Measure the resonant frequency and the -3 dB bandwidth. Repeat for $R = 200\ \Omega$, and plot V_L, V_C, V_R, and the circuit impedance "seen" by the source.

6.8   SERIES-TUNED LCR PHASE RESPONSE

Fig. 6.6 shows a current phase marker placed across the 20 Ω resistor to plot the phase response.

Be very careful about placing and rotating components on the screen, as it may produce unexpected results. With the cursor icon selected, pressing the ⊞ icon makes the cursor jump to the middle of the phase response. Pressing the ☒ icon places x and y signal values at this point. For very high Q values, the slope of the phase response is very steep around resonance, so you might not be able to measure accurately the phase at resonance. One solution is to increase the number of points plotted and/or decrease the frequency range around resonance. Press **F11** to obtain the phase response in Fig. 6.7.

Select the magnifying glass icon to examine the region around the resonant point. If the response is not smooth, repeat the analysis but with a reduced frequency range and an increased number of points. Measure the frequency on the phase response at $-45°$, $0°$, and $+45°$ and calculate the bandwidth. **Lclick** the mouse button for the first cursor and move the mouse,

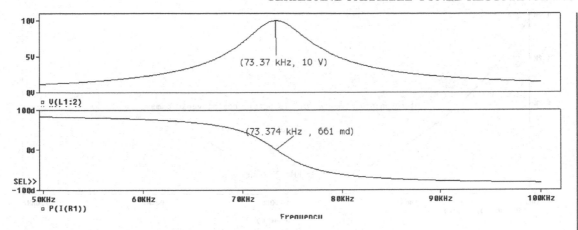

FIGURE 6.7: Current phase response

then click the right mouse button for the second cursor. The cursors may be moved to find the −3 dB frequency values (If the display is not in dB, then measure the frequency where the voltage is reduced by 70.7% of the maximum voltage). For example, if the maximum voltage is 10 V, then measure the frequencies f_1 and f_2 on either side of the resonant frequency where the voltage is reduced to 7.07 V. Dividing f_0 by the BW to calculate the loaded Q-factor. From the phase response, compare measured and calculated bandwidths.

6.9 IMPEDANCE OF A SERIES-TUNED CIRCUIT

To plot the circuit impedance, run a simulation without markers placed on the circuit. From the Probe screen, click the **Trace** menu (or press the **insert** key), and add a trace as in Fig. 6.8. From the list, select **V(VIN)** and insert "/" in the **Trace Expression** box at the bottom left side of the screen end and then select **I(Vin)**.

The impedance of the tuned circuit $Z = V(source)/I(Vin)$ is plotted in Fig. 6.9 with the x-axis set to log. A minimum value occurs at the resonant frequency and is the resistance of the inductor.

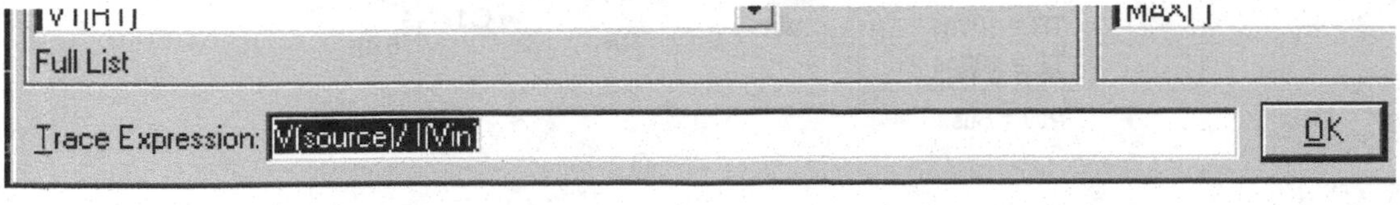

FIGURE 6.8: Trace expression box

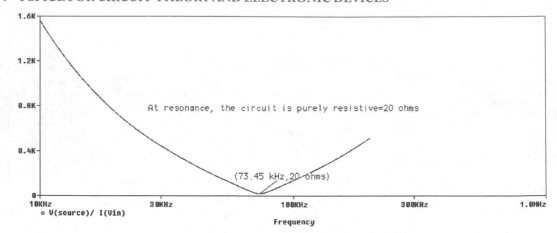

FIGURE 6.9: Series impedance plot

6.10   FOURIER SERIES

A *periodic* function, $f(t)$, with period T and fundamental frequency, f_0, may be generated using the sum of N weighted cosine and sine components using the Fourier series expansion (Joseph Fourier 1768–1830):

$$f(t) = V_0 + \sum_{n=1}^{N} (a_n \cos 2\pi n f_0 t + b_n \sin 2\pi n f_0 t) \qquad (6.20)$$

V_0 is a DC term, f_0 is the fundamental frequency of the signal and weighting coefficients a_n and b_n. Thus, a square wave may be broken into a series of harmonic terms and represented by the equation in (6.20) [ref: 6].

6.10.1   Series-Tuned Circuit as a Low-Pass Filter

A 100 kHz square wave applied to the series-tuned resonant circuit in Fig. 6.10 has the fundamental component of the 100 kHz square wave extracted by the low-pass filter.

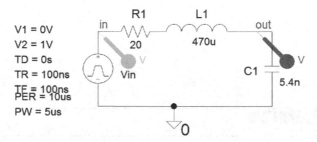

FIGURE 6.10: Second-order LPF

TABLE 6.1: VPULSE Parameters
DC = 0 (offset)
AC = 0.1 (amplitude)
V1 = 0 (space value)
V2 = 1 (mark value)
TD = 0 s (delay time)
TR = 0.1 us (rise time)
TF = 0.1 us (fall time)
PW = 5 us (pulse width)
PER = 10 us (pulse period)

Place a **VPULSE** generator from the **source.olb** library and set the parameters as shown in Table 6.1. Enter numerical values only (The parameter description is for your information only). The pulse period **PER** $T_p = 10$ us is the inverse of the resonant frequency $f_0 = 100$ kHz and the pulse width **PW** $= 5$ us (for an equal mark-space ratio).

6.10.2 The Output File

The junction at L and C is labeled **out**. Select the analysis set-up icon and set the parameters as shown. Select the **Output File Options** as shown in Fig. 6.11, and tick the **Perform Fourier Analysis** parameters. The two variable entered in **Output Vars** box are: **V(in) V(out)**, where in and out are the wire segment names separated by a space. The input and output spectral information will be in the output file accessed from the **PSpice/View Output File** menu.

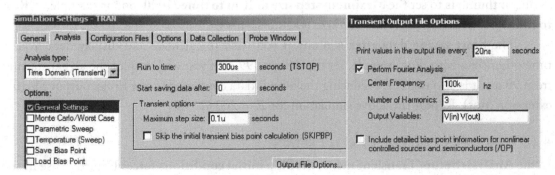

FIGURE 6.11: Transient set-up

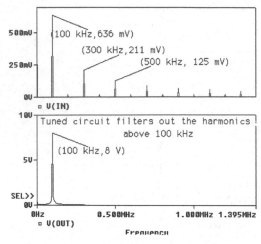 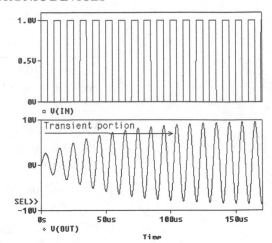

FIGURE 6.12: Input and output signals

Press **F11** to simulate. The plot shown in Fig. 6.12 is for $R = 20\ \Omega$.

The extra plots were obtained pressing **Alt PP** keys and copying the input and output variables onto the extra plots. From Probe, select **Plot/Unsynchronize X Axis menu** to change the horizontal axis to frequency. Press the **FFT** icon to display the two plots in the frequency domain. The extracted 100 kHz sine wave signal has a period of 10 us. The step size is automatically adjusted by the PSpice dynamic internal algorithm step size change, increasing its size during periods when there is little activity, but decreasing the step size for rapid time changing events. Too large a **Maximum step size** value makes the display too course and angular in shape. The ratio **Run to time/Maximum step size** specifies the number of points plotted. For example, 100 points are plotted if the ratio is 100 us/1 us. A small **Maximum step size** value will make the analysis time longer but it will produce a smoother-looking plot (reduce this parameter if you have convergence problems). If you observe triangular-shaped signal waveforms, as in Fig. 6.13, it means the maximum step size chosen is not small enough. A rule of thumb is to set the **Maximum step size** to **Run to time /1000**. So for example, if **Run to time = 15 s**, then **Maximum step size** is set to 15 ms.

The **Start saving data after** is useful if you wish to calculate and display data from a time other than zero. For example, set the delay to 99 us if you wish to plot the signal starting from 99 us. It is also useful for reducing the output data contained in the Probe output file (file extension **.out**). Repeat the above simulation for $R = 200\ \Omega$.

6.11 SKIP INITIAL CONDITIONS

In the **Analysis Setup/Transient** menu, the **Skip initial conditions** forces PSpice to use initial conditions on an inductor or capacitor. If not checked, PSpice calculates the initial conditions

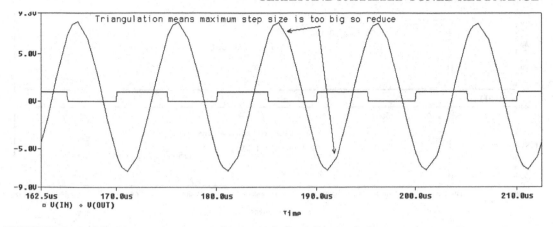

FIGURE 6.13: Maximum step size too large results in a distorted plot

treating capacitors as open circuits and inductors as short circuits. Care must be taken to ensure that the **Skip initial conditions** box is unchecked in schematics where initial conditions are required. The Fast Fourier Transform icon (**FFT**) in Probe is useful for examining the frequency components of a complex signal, as demonstrated in Fig. 6.14.

Increase the **Run to time** and/or decrease the **Maximum step size** to increase the **FFT** resolution. The spectrum of the output signal displayed in the top plot shows higher spectral components almost eliminated by the filtering action of the circuit.

6.12 PARALLEL-TUNED *LCR* CIRCUIT

The parallel-tuned circuit in Fig. 6.15 shows a capacitor in parallel with an inductor whose series coil resistance is r_c. The circuit is fed from a current source comprising an ideal current source in parallel with a large source resistance value, hence the source resistance will produce

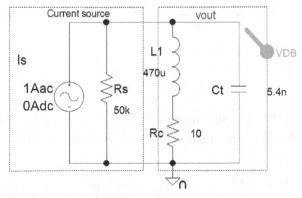

FIGURE 6.14: Spectrum of a square wave

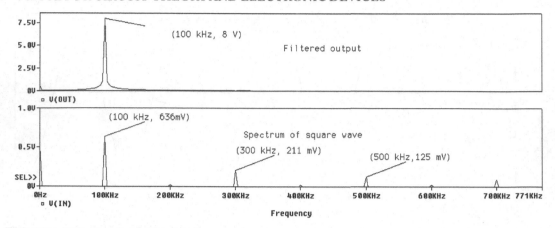

FIGURE 6.15: Parallel LCR circuit

minimum loading. *Note*: In PSpice, the voltage and current sources are ideal, hence a current source resistance is an open circuit, whilst a voltage source resistance is a short circuit. We need to add a resistance across the current source in order to simulate, otherwise an error will occur. Likewise, it is good practice to add a source resistance in series with a voltage source. Sometimes it is necessary to represent a current source as a voltage source in series with a high resistance.

It is easier to analyze this circuit if we replace the coil resistance and reactance of the series branching $Z = R_c + j\omega L$ with an equivalent parallel *admittance* consisting of a *conductance* **G** and *susceptance* **B**, i.e. $Y = G + jB$.

$$Y = \frac{1}{Z} = \frac{1}{R_c + j\omega L} \times \frac{R_c - j\omega L}{R_c - j\omega L} = \frac{R_c}{R_c^2 + \omega^2 L^2} - j\frac{\omega L}{R_c^2 + \omega^2 L^2} = G + jB \qquad (6.21)$$

Inverting the conductance G gives the equivalent parallel resistance

$$R_P = \frac{1}{G} = \frac{R_c^2 + \omega^2 L^2}{R_c} = R_c + \frac{\omega^2 L^2}{R_c} = R_c + \frac{\omega^2 L^2}{R_c}\left(\frac{\times R_c}{\times R_c}\right) = R_c + \frac{\omega^2 L^2}{R_c^2}R_c \qquad (6.22)$$

Introduce the Q-factor $Q_0 = \frac{\omega L}{R_c}$ into (6.22)

$$R_P = R_c + Q_0^2 R_c \qquad (6.23)$$

This resistance is called the dynamic impedance of the circuit at resonance (it is resistive at resonance). We can make the approximation $R_P \approx Q_0^2 R_c$, and is valid for $Q_0 > 10$. The parallel reactance is

$$X_{LP} = \frac{1}{B} = \frac{R_c^2 + \omega^2 L^2}{\omega L} = \omega L + \frac{R_c^2}{\omega L}\left(\frac{\times \omega L}{\times \omega L}\right) = \omega L + \left(\frac{R_c}{\omega L}\right)^2 \omega L \qquad (6.24)$$

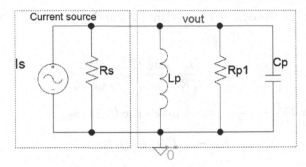

FIGURE 6.16: Equivalent parallel LCR circuit

Substituting for the unloaded Q-factor yields

$$X_{LP} = \omega L + \frac{\omega L}{Q^2} \qquad (6.25)$$

The circuit is now represented by the equivalent circuit shown in Fig. 6.16.

For Q values greater than 10, we make the approximation that the parallel inductive reactance is equal to the series inductive reactance i.e. $X_{LP} \approx X_{LS} = \omega L$. At resonance, the total reactance is zero, so $X_{LP} = X_{LS} = X_C$.

$$\frac{R_c^2 + \omega_0^2 L^2}{\omega_0 L} = \frac{1}{\omega_0 C} \Rightarrow R_c^2 + \omega_0^2 L^2 = L/C \qquad (6.26)$$

Rearranging (6.26) gives an expression for the frequency of resonance as

$$\omega_0^2 = \frac{1}{L^2}\left[\frac{L}{C} - R_c^2\right] = \frac{1}{LC} - \frac{R_c^2}{L^2} \qquad (6.27)$$

Expressed in hertz:

$$f_0 = \frac{1}{2\pi}\sqrt{\frac{1}{LC} - \frac{R_c^2}{L^2}} \qquad (6.28)$$

From (6.22) we can write

$$G = \frac{R_c}{R_c^2 + \omega^2 L^2} = \frac{1}{R_P} \qquad (6.29)$$

Replacing $R_c^2 + \omega^2 L^2$ with L/C from (6.26) yields the dynamic impedance as

$$R_P = \frac{L}{R_c C} \qquad (6.30)$$

To express Q in terms of the dynamic impedance R_p and X_L, multiply (6.30) by the resonant frequency as

$$R_P = \frac{L}{CR_c}\left(\frac{\omega_0}{\omega_0}\right) = \frac{\omega_0 L}{\omega_0 C R_c} \tag{6.31}$$

The Q-factor is $Q_0 = \frac{\omega_0 L}{R_c} = \frac{1}{\omega_0 C R_c}$, so we may write (6.31) as

$$R_P = \omega_0 L Q_0 = Q_0/\omega_0 C \tag{6.32}$$

6.12.1 Universal Response Curve

The normalized impedance of a parallel-tuned circuit is expressed in terms of the Q-factor and the fractional de-tuning factor, x. The impedance of a parallel-tuned circuit is

$$Z = \frac{(R_c + jX_L)(-jX_C)}{R_c + jX_L - jX_C} \tag{6.33}$$

We simplify this expression by assuming a high Q-factor coil, so that $X_L \gg r_c$

$$Z = \frac{X_C X_L}{R_c + j(X_L - X_C)} = \frac{L/C}{R_c + j(X_L - X_C)} \tag{6.34}$$

Dividing (6.34) above and below by r_c

$$Z = \frac{L/CR_c}{1 + j1/R_c(X_L - X_C)} \tag{6.35}$$

The impedance at resonance is called the dynamic impedance $R_P = L/CR_c$, even though it is resistive. Divide (6.35) by the dynamic impedance to normalize as

$$\frac{Z}{R_P} = \frac{1}{1 + j1/R_c(X_L - X_C)} = \frac{1}{1 + j\omega L/R_c(1 - 1/\omega^2 LC)} = \frac{1}{1 + j\omega L/R_c(1 - \omega_0^2/\omega^2)} \tag{6.36}$$

$$\frac{Z}{R_P} = \frac{1}{1 + j\frac{\omega L}{R_c} \times \frac{\omega_0}{\omega_0}(1 - \frac{\omega_0^2}{\omega^2})} = \frac{1}{1 + j\frac{\omega L}{R_c} \times \frac{\omega_0 L}{R_c}(\frac{\omega}{\omega_0} - \frac{\omega_0}{\omega})} = \frac{1}{1 + jQ_0\left(\frac{\omega}{\omega_0} - \frac{\omega_0}{\omega}\right)} \tag{6.37}$$

The fractional de-tuning factor, x, is

$$x = \frac{\omega_0 - \omega}{\omega_0} = 1 - \frac{\omega}{\omega_0} \Rightarrow \frac{\omega}{\omega_0} = 1 - x \tag{6.38}$$

So that $\frac{\omega}{\omega_0} = 1 - x \Rightarrow \frac{\omega_0}{\omega} = \frac{1}{1-x}$. Combine these two equations yields

$$\frac{\omega}{\omega_0} - \frac{\omega_0}{\omega} = \frac{1-x}{1} - \frac{1}{1-x} = \frac{1 - 2x + x^2 - 1}{1 - x} = \frac{x^2 - 2x}{1 - x} \tag{6.39}$$

Substitute (6.39) into (6.37) and express the magnitude of the result as

$$\left|\frac{Z}{R_P}\right| = \frac{1}{\sqrt{1 + Q_0^2 \left(\frac{\omega}{\omega_0} - \frac{\omega_0}{\omega}\right)^2}} = \frac{1}{\sqrt{1 + Q_0^2 \left(\frac{x^2 - 2x}{1 - x}\right)^2}} \qquad (6.40)$$

For $x \ll 1$, i.e. frequencies close to f_0 we can approximate (6.40) as

$$\left|\frac{Z}{R_P}\right| \approx \frac{1}{\sqrt{1 + 4Q_0^2 x^2}} \qquad (6.41)$$

Expressed in terms of the fractional de-tuning, α

$$\left|\frac{Z}{R_P}\right| = \frac{1}{\sqrt{(1 + 4Q_0^2 \alpha^2}} \qquad (6.42)$$

6.12.2 Relationship Between the Resonant Frequency and Bandwidth

A similar result for the series-tuned normalized current is $|i/i_0| = 1/\sqrt{1 + 4Q_0^2 \alpha^2}$, hence the normalized power is

$$\frac{i^2 R}{i_0^2 R} = \frac{P}{P_0} = \frac{1}{1 + 4Q_0^2 x^2} \qquad (6.43)$$

The -3 dB frequency is where the maximum output power P_0 falls to half of its value.

$$1 + 4Q_0^2 x^2 = 2 \Rightarrow 4Q_0^2 x^2 = 1$$

$$4Q_0^2 x^2 = 1 \Rightarrow x^2 = \frac{1}{4Q_0^2} \Rightarrow x = \pm\frac{1}{2Q_0}$$

Defining x as

$$x = \frac{\omega_0 - \omega}{\omega_0} = \frac{\Delta\omega}{\omega_0} = \frac{\Delta f}{f_0} = \pm\frac{1}{2Q_0} \qquad (6.44)$$

Since the bandwidth is $2\Delta f$, we may write

$$\frac{2\Delta f}{f_0} = \frac{BW}{f_0} = \frac{1}{Q_0} \Rightarrow Q_0 = \frac{f_0}{BW} \qquad (6.45)$$

6.12.3 Loaded Q-factor

A parallel-tuned circuit is connected to a nonideal current source with a high impedance source. The source will load the tuned circuit because any extra resistance in parallel reduces the Q-factor. Because of this, we must introduce the loaded Q-factor. The equivalent parallel resistance at resonance is the dynamic impedance R_p and is reduced by any external resistance R_{EXT} placed in parallel with the circuit. The loaded Q-factor is $Q_L = R_T/X_L$, with the total resistive loading $R_T = R_S//R_P//R_{EXT} = \frac{RR_P}{R+R_P} = \frac{R_P}{1+R_P/R}$, and $R = R_S R_{EXT}/(R_S + R_{EXT})$.

6.13   EXAMPLE

The drain circuit of a FET tuned radio frequency amplifier has a 100 pF capacitor placed in parallel with an inductor L, whose unloaded Q-factor is 100. If the frequency of resonance is 1 MHz and the transistor output resistance is 20 kΩ, calculate the loaded Q-factor, inductance, and loaded bandwidth.

6.13.1 Solution

The total resistive loading on the tuned circuit consists of the transistor output impedance and the dynamic impedance in parallel $R = R_S R_P/(R_S + R_P)$. The dynamic impedance using the unloaded Q-factor is

$$R_P = \frac{Q_{UL}}{\omega_0 C} = \frac{100}{2\pi(1 \times 10^6)(100 \times 10^{-12})} = \frac{10^6}{2\pi}\Omega = 159 \text{ k}\Omega$$

Substitute the two impedances, gives the total resistive loading $R = \frac{20.10^3 \times 159.10^3}{20.10^3 + 159.10^3} = 17.76$ kΩ. Since the unloaded Q-factor is greater than 10, then the dynamic impedance $R_P \approx Q_{UL}^2 R_c$, hence the coil resistance is $159 \times 10^3/(100)^2 = R_c = 15.9$ Ω. The loaded Q-factor $Q_L = \frac{Q_{UI}}{1+R_P/R} = \frac{100}{1+159k/17.76k} = 10$ and the unloaded Q-factor is $Q_{UL} = \omega_0 L/R_c \Rightarrow L = Q_{UL} R_c/\omega_0 = 100(15 \cdot 9)/2\pi(1 \times 10^6) = 0 \cdot 253$ mH. The unloaded bandwidth is $f_0/Q_{UL} = 1$ MHz/100 $= 10$ kHz, and the loaded bandwidth is 1 MHz/10 k $= 100$ kHz [ref: 3].

6.14   PROBLEM

For the circuit in Fig. 6.17, calculate the unloaded and loaded Q-factors, resonant frequency, the -3 dB bandwidth (BW) and the dynamic impedance. Compare these calculations to the simulate resultants from the voltage amplitude response. Plot the phase and impedance responses across the circuit (do not include the source resistance).

An ideal voltage source resistance is zero, so connecting it across a parallel LCR circuit would destroy the circuit selectivity. The circuit must therefore be fed from a current source (An ideal current source has an infinite source resistance) and here we mimic a current source as a voltage source in series with a high source resistance. Set the **Analysis** tab to **Analysis type: AC Sweep/Noise, AC Sweep Type to Logarithmic, Start Frequency = 100, End Frequency = 1000k, Points/Decade = 1000.** In general, to set the frequency range you need to calculate

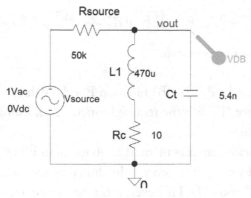

FIGURE 6.17: Thévenin equivalent circuit for a current source

the resonant frequency first. The **Start Frequency** and **End Frequency** values are then set a decade above and below the resonant frequency and the *x*-axis set to **Linear**.

6.15   FREQUENCY RESPONSE OF A PARALLEL-TUNED CIRCUIT
The parallel-tuned frequency response is shown in Fig. 6.18.

Use the cursors and magnifying tool to measure the −3 dB bandwidth and resonant frequency accurately. If the plot looks triangular, reset the frequency range from 90 k to 120 k and simulate. Remember using the magnifying tool does not increase the number of plotted points, so you need to reduce the frequency range and/or the number of points plotted. From **Probe/Trace/Evaluate measurement** menu, press the sixth icon shown in Fig. 6.19.

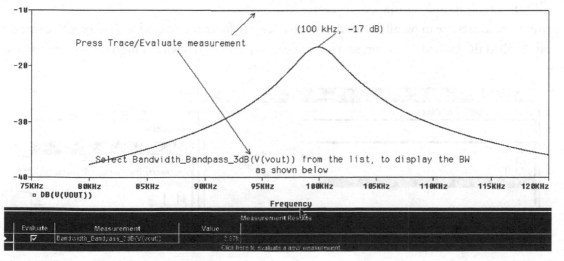

FIGURE 6.18: Output voltage in dB

FIGURE 6.19: Evaluate measurement icon

Select the -3 dB bandwidth (1 dB) function **Bandwidth (1, dB)** function from the list shown in Fig. 6.20. Replace "1" with the required function i.e. **Bandwidth(V(Ct:2), 3)**, where 3 represents the -3 dB point.

From the **Trace/Measurements** menu the sub menu in Fig. 6.21 appears. Pressing **Eval** opens a further menu and you need to browse the directory and insert the function **V(Ct:2)** to display the bandwidth as 3.966 kHz. However, this information disappears when you press ok.

Place a phase marker from the **PSpice/Markers/Advanced** menu to observe the phase response shown in Fig. 6.22.

The phase at the resonant frequency is zero, but why is the phase of the voltage and current zero at DC (Hint: Assume the capacitor reactance is infinite and the inductive reactance is zero at very low frequencies)?

6.16   DYNAMIC IMPEDANCE

To plot the impedance of the circuit, press the insert button on the keyboard, and in the **Trace Expression** box shown in Fig. 6.23, insert **V1(Vsource)/ I(Vsource)**.

The circuit impedance $Z = $ **V1(Vsource)/I(Vsource)** is shown in Fig. 6.24. Measure the impedance at the resonant frequency and subtract the 50 kΩ source resistance and verify the dynamic impedance is $(R_p = L1/C_t r_c) = 8.7$ kΩ.

6.16.1 Loaded and Unloaded Q-factor

Any extra resistance in parallel with the tuned circuit reduces the unloaded Q-factor to a smaller value called the loaded Q-factor, so a current source, although having a high source impedance,

FIGURE 6.20: The -3 dB BW evaluate measurement function

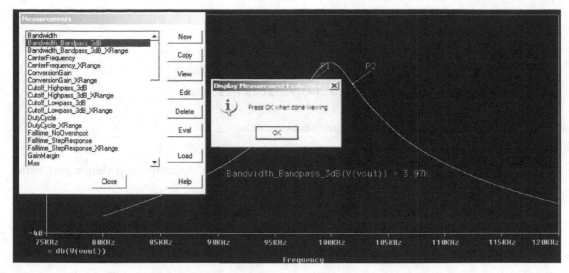

FIGURE 6.21: Measurement function display

still reduces the dynamic impedance and hence the Q-factor. To determine the Q-factor, the dynamic resistance R_p, and the AC coil resistance Rc, draw the schematic in Fig. 6.25. Set the **VSIN** generator to 10 V at a resonant frequency of 100 kHz. Measure the output voltage across the capacitor. From potential division, and the measured values input and output voltages V_s and V_{out}, calculate the dynamic impedance R_p, the AC coil resistance R_c, and Q_{UL}.

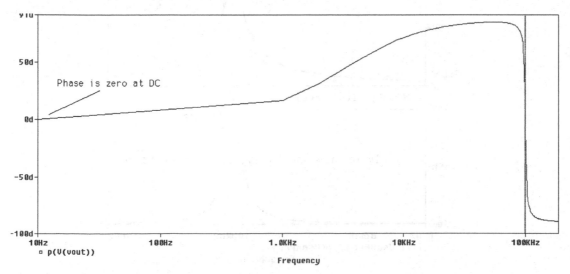

FIGURE 6.22: Phase relationship for the capacitor voltage

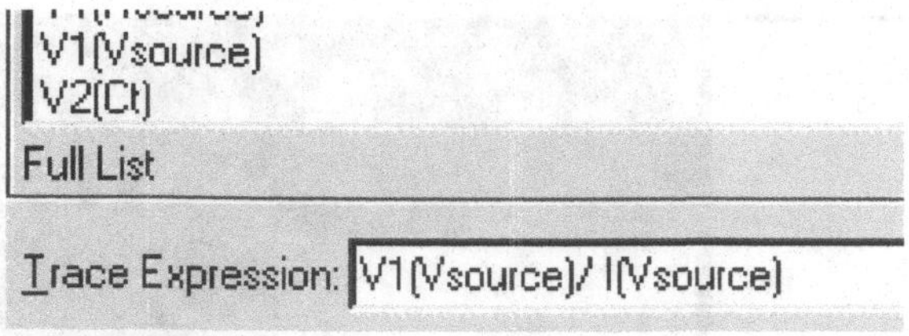

V1(Vsource)
V2(Ct)

Full List

Trace Expression: V1(Vsource)/ I(Vsource)

FIGURE 6.23: Adding the source voltage/source current

The unloaded Q-factor is expressed in terms of the dynamic impedance and series reactance as

$$Q_{UL} = \frac{R_P}{X_L} = \frac{R_P}{\omega_0 L} \approx \frac{R_P}{\frac{1}{\sqrt{LC}}L} \approx \frac{R_P}{\sqrt{\frac{L^2}{LC}}} = R_P\sqrt{\frac{C}{L}} \qquad (6.46)$$

R_P is the *dynamic impedance* and is the parallel resistance of the unloaded tuned circuit at resonance. Any external resistance in parallel reduces this resistance and changes the unloaded Q-factor to a loaded Q-factor thus increasing the bandwidth and reducing selectivity. If the total resistance is R in parallel with R_p, then the loaded Q-factor is

$$Q_L = R_T/X_L = \frac{R_P R/(R_P + R)}{X_L} \qquad (6.47)$$

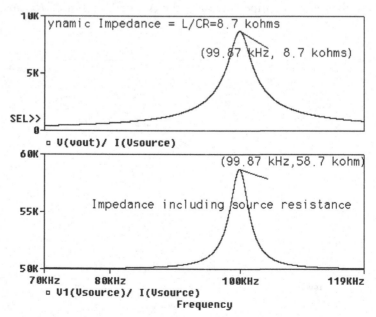

FIGURE 6.24: Plotting the circuit impedance

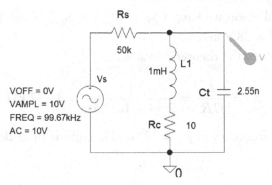

FIGURE 6.25: Q-factor measurement

R is the source resistance in parallel with external resistive loading R_{EXT}, hence the total resistance is $R = R_S R_{\mathrm{EXT}}/(R_S + R_{\mathrm{EXT}})$. The loaded Q in terms of the unloaded Q-factor is

$$Q_L = \frac{R_P}{X_L}\frac{1}{1 + R_P/R} = \frac{Q_{UL}}{1 + R_P/R} \qquad (6.48)$$

The Q-factor increases as the C/L ratio increases and is the opposite result to that obtained for the series-tuned circuit. Simulate to produce the result in Fig. 6.26. The output voltage is calculated by potential division at f_0, as 4.39 V for an input $V_{\mathrm{in}} = 10$ V:

$$V_{\mathrm{out}} = V_{\mathrm{in}}\frac{R_P}{R_P + R_s} = \frac{V_{\mathrm{in}}}{1 + R_s/R_P} \qquad (6.49)$$

R_s is the source resistance and the dynamic impedance $R_P = L/C R_c$ (R_c is the ac coil resistance and is not easily measured but assume 10 Ω). If the measured voltage across the circuit is 4.39

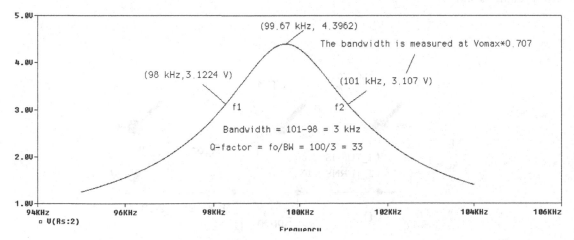

FIGURE 6.26: Measure the voltage at f_0

V, then from potential division we have $4.38 = 10/(1 + R_s/R_P) \Rightarrow R_s/R_P = 1.28 \Rightarrow R_P = R_s/1.28 = 50.10^3/1.28 = 38.9 \text{ k}\Omega$.

Compare this value to the theoretical value:

$$R_P = \frac{L}{CR_c} = \frac{10^{-3}}{2.54 \times 10^{-9}.10} = 39.37 \text{ k}\Omega \qquad (6.50)$$

The measured resonant frequency is $f_0 = 99.86$ kHz and the unloaded Q-factor is calculated as

$$Q_{UL} = R_P/X_{LP} \approx R_P/X_{LS} = 38{,}900/(2\pi f_0 L) = 38{,}900/(2\pi 99.8) = 62$$

The loaded Q-factor $= R/X_{LP}$, is approximately equal to R/X_{LS}, where R is R_p in parallel with R_s equal to 21.87 kΩ.

6.17 TRANSFORMERS

William Stanley (1858–1916) built an induction coil that was the precursor to the transformer. A transformer has primary and secondary coils magnetically coupled, with each winding wound on the same former (core) made from materials such as iron, steel but RF transformers generally have an air core. All PSpice schematics require a ground part so we have to attach a second ground symbol to the secondary or a very large resistance between primary and secondary as shown in Fig. 6.27, otherwise we get a floating error as it "sees" the secondary circuit floating.

This is a useful technique whenever a floating error occurs, for example, if the output component is a capacitor then one end is seen as floating. The transformer used here is a **K3019PL_3C8** part in the **eval.olb** library and the new power markers are located on the component as shown.

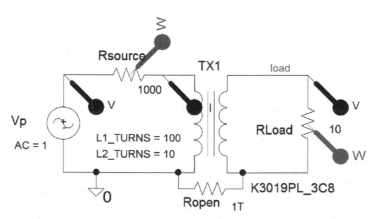

FIGURE 6.27: Transformer

COUPLING	0.999
L1_TURNS	100
L2_TURNS	10
Location X-Coordinate	680
Location Y-Coordinate	495
Source Part	K3019PL_3C8.Normal

FIGURE 6.28: Transformer parameters

6.17.1 Transformer Parameters

Other transformers are: K502T300_3C8, K528T500_3C8, K_Linear, XFRM_LINEAR and XFRM_NONLINEAR. Edit the properties by **Rclicking** and selecting **Edit Properties** to enter the parameters as shown in Fig. 6.28.

Transformer parts such as the **XFRM_LINEAR** part require coupling and inductance values.

6.17.2 Matching Transformer

Maximum power transfer is transferred from a source to a load only when they are equal. A transformer can "match" a load to a source and to show this we need to derive expressions for the primary and secondary parameters in terms of the ratio of the primary and secondary turns-ratio. To simplify the analysis, assume the power in the primary circuit is equal to the power in the secondary circuit—a valid approximation because transformers are very efficient and dissipate little power, hence we may write

$$v_P i_P = i_S v_S \Rightarrow v_P = \frac{i_P}{i_S} v_S \qquad (6.51)$$

Equate the primary ampere-turns to the secondary ampere-turns as

$$i_P N_P = i_S N_S \Rightarrow \frac{i_P}{i_S} = \frac{N_S}{N_P} = n \Rightarrow i_P = \frac{i_S}{n} \qquad (6.52)$$

Substituting (6.52) into (6.51) yields

$$v_P = n v_S \qquad (6.53)$$

The primary impedance is the primary voltage divided by the primary current. Substituting for the primary current and voltage from (6.52) and (6.53) yields

$$Z_P = \frac{v_P}{i_P} = \frac{n v_S}{i_{S/n}} \Rightarrow Z_P = n^2 Z_S \qquad (6.54)$$

Fig. 6.29 shows the input and output power dissipated in the source and load resistances.

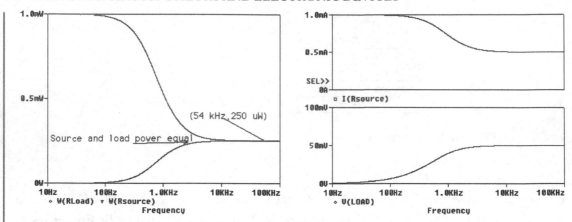

FIGURE 6.29: Matching transformer for maximum power transfer

The primary circuit "see" a much higher resistance i.e. n^2 times the load resistance. For $n^2 = 100$, the primary circuit is "looking" at 1000 Ω and not 10 Ω. Any resistance connected to the secondary side will appear in the primary side of the transformer, as that resistance multiplied by the square of the turns-ratio n. A multi-winding transformer may also be created using the **K_Linear** part that links together individual inductor coil names.

6.18 POWER SUPPLIES: RECTIFICATION AND REGULATION

The schematic in Fig. 6.30 shows the input 220 V RMS mains voltage (311 V peak) with a mains frequency of 50 Hz for European electricity supplies but is 110 V/60 Hz in the United States. The 0.001 Ω mains source resistance is small since it is a voltage source and R5 is a large terra-ohm (1 TΩ) resistance to avoid a floating error. Differential voltage markers from the **Markers** menu are placed across the secondary terminals as shown. These are necessary since the measurement reference is not with respect to the ground. Be sure to remove $R5$ if you are creating a printed circuit.

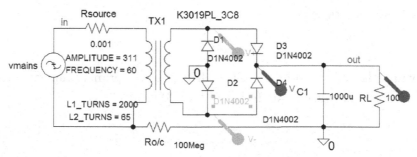

FIGURE 6.30: Power supply

COUPLING	0.999
L1_TURNS	2000
L2_TURNS	65
Location X-Coordinate	680
Location Y-Coordinate	495
Source Part	K3019PL_3C8.Normal

FIGURE 6.31: Setting the turns-ratio for the mains transformer

The ripple voltage is calculated

$$V_{\text{peak ripple}} = \sqrt{3}V_{r(\text{RMS})} = \sqrt{3}(2.4I_{\text{DC}}/C) \qquad (6.55)$$

The DC voltage across C is

$$V_{\text{DC}} = V_{\text{sec}} - V_{\text{peak ripple}} \qquad (6.56)$$

6.18.1 Power Supply Waveforms

Select the properties by **Rclicking** and entering the transformer parameters: **coupling** = 0.999, **L1_TURNS** = 2000 and **L2_TURNS** = 65. The primary voltage peak is 311 V, and the secondary peak is 10 V—a turns-ratio of 30:1, as shown in Fig. 6.31.

Set the transient analysis to display a few cycles after simulation as shown in Fig. 6.32. Verify using (6.55) and (6.56) that all voltage waveforms have the correct values.

Fig. 6.33 shows a display of the RMS and instantaneous power in the load.

6.18.2 Power Supply Voltage Regulation

The ripple voltage shown in Fig. 6.34 may manifest itself in audio power amplifiers as an audible 120 Hz mains hum (remember it is a rectified sine wave so it is twice the mains frequency supply). To eliminate ripple we need to regulate the power supply using a voltage regulator. The 15 V regulated power supply has a larger number of secondary winding turns to increase the secondary output voltage in order for the LM7815 regulator to work correctly. This voltage regulator is not part of the evaluation version so we use the symbol from **mylib.olb** and attach to it the LM7815c model downloaded from the Internet.

The regulated voltage is clearly seen from Fig. 6.35 where the ripple voltage is reduced.

6.19 GROUND BOUNCE

The schematic in Fig. 6.36 investigates **"ground bounce"**. PSpice hides digital IC power supply pins but we may connect power supplies to the hidden pins by **DLclicking** the IC and ticking **Power Pins Visible**.

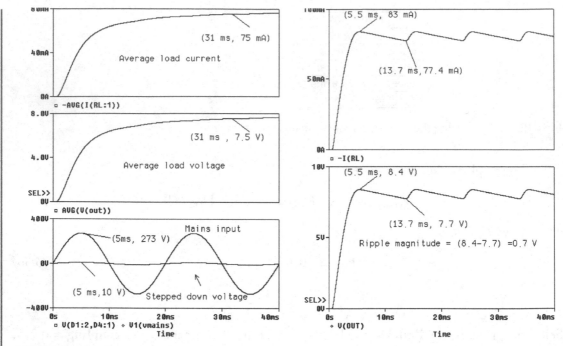

FIGURE 6.32: Power supply waveforms

Decouple each IC by placing a 100 nF capacitor close to the IC supply pin. In digital circuits, fast rise-time signals produce problems in the DC supply lines because PCB power supply tracks and IC pins possess inductance and hence, together with track capacitance, produces oscillatory signals on the DC rails. Select the 7474 IC D flip-flop part, **RClick/Edit**

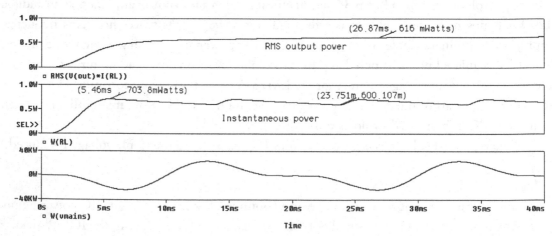

FIGURE 6.33: Output power

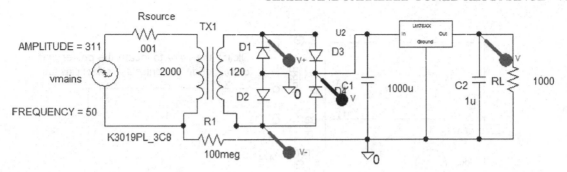

FIGURE 6.34: Voltage regulation

Properties, to show the properties spreadsheet. In the spreadsheet tick **Power Pins Visible** as shown in Fig. 6.37.

The power supply connections, normally hidden, have inductance and capacitance added to mimic the effects of PCB and IC output capacitance and inductance and allow us to investigate how these components produce oscillatory signals on the power supply line rails. Observe ground bounce at the bottom of Fig. 6.38.

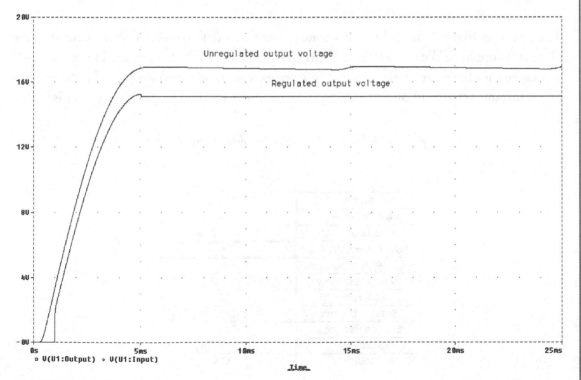

FIGURE 6.35: Regulated output voltage

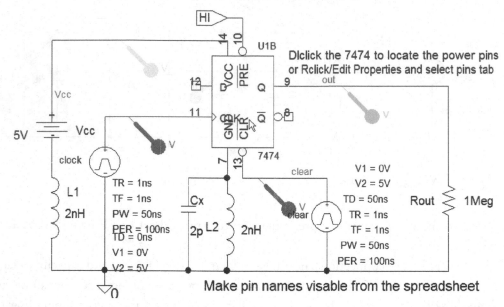

FIGURE 6.36: Ground bounce

6.20   POWER FACTOR CORRECTION

The circuit in Fig. 6.39 is used to investigate power factor (PF) correction as demanded by the electricity suppliers. We use the **Trace Expression** facility in Probe to plot certain mathematical expressions for the power in the circuit. The instantaneous power in a passive load is the product of the instantaneous load current and load voltage. For a sinusoidal signal, the power is

$$p(t) = v(t)i(t) = V_m \cos(\omega t + \phi) I_m \cos(\omega t + \theta) \tag{6.57}$$

Name	INS10005
Part Reference	U2A
PCB Footprint	DIP.100/14/W.300/L.800
Power Pins Visible	☑
Primitive	DEFAULT
PSpiceTemplate	X^@REFDES %C\R\ %D %
Reference	U2
Source Library	C:\ORCAD\ORCAD_10.
Source Package	7474
Source Part	7474.Normal
Value	7474

FIGURE 6.37: The 7474 spreadsheet

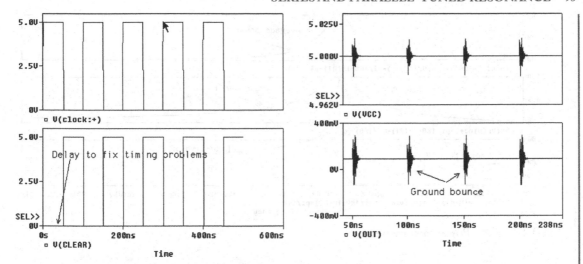

FIGURE 6.38: Ground bounce waveforms

Expanding (6.57) yields

$$p(t) = 0.5 V_m I_m \cos\theta + 0.5 V_m I_m \cos(2\omega t + \phi + \theta) \qquad (6.58)$$

The first part of (6.57) is the average power consumed and charged for, by the electricity suppliers. Power factor angle is zero for resistive loads and the average power is 0.5 $V_m I_m = V_{RMS} I_{RMS}$ W. A lagging PF $= \cos\varphi = \phi - \theta$ is positive for inductive loads, and a negative leading PF for capacitive loads, as shown in Fig. 6.40. To plot degrees in PSpice, we must multiply the angle in radians by pi/180 ($= 1/57.3$), i.e.

$$V(vin : +)^*I(R1)^*\cos(VP(vin : +)^*pi/180 - IP(R1)^*pi/180)$$

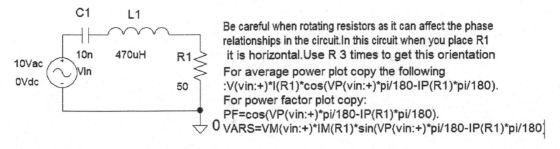

FIGURE 6.39: Power factor correction

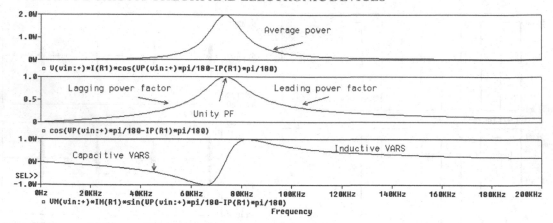

FIGURE 6.40: Power factor waveforms

6.20.1 Average Power and Apparent Power

As the power factor angle increases, the average power decreases, but the *peak load current* remains unchanged producing increased resistive losses which increases the costs to the electrical suppliers. They, in turn, charge the consumer for producing nonunity power factors. The apparent power is defined as

$$0.5 V_m I_m = \frac{p(t)}{\cos\theta} = \frac{p(t)}{\text{PF}} \text{ VA} \qquad (6.59)$$

Fig. 6.40 shows the power factor varying from zero, for a pure reactive load, to unity for a resistive load. The second term in (6.57) has a zero average value and is the power stored in the reactive elements of the load and returned to the source. When positive, the total power is greater than the average power and is absorbed by the load. When negative, the total power is less than the average power. When power factors are less than 0.5, the peak value of the second

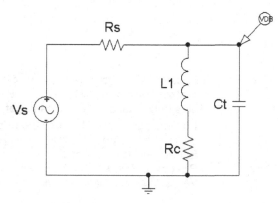

FIGURE 6.41: Parallel-tuned LCR circuit

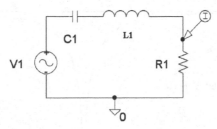

FIGURE 6.42: Frequency response

term exceeds the average power, and on negative peaks, power flows from the load back to the source. Reactive power is measured in **volt-amp reactive (VARS)** and is $Q = 0.5 V_m I_m \sin \theta$. Complex power, S, has dimensions of *volt-amp (VA)* and is related to average power P and the peak reactive power, Q, as $S = P + jQ$, where P and Q are the resistive and reactive load components defined: $P = 0.5 I_m^2 R$ and $Q = 0.5 I_m^2 X$.

6.21 EXERCISES

(1) Fig. 6.41 shows a resonant circuit fed from a voltage source in series with a large resistance Rs thus simulating a current source. Determine the -3 dB bandwidth, loaded and unloaded Q-factor, and the dynamic impedance $R_s = 110 \, k\Omega$, $R_c = 10 \, \Omega$, $Ct = 5.3 \, nF$, $L_1 = 470 \, uH$,

(2) The parallel-tuned frequency response is shown in Fig. 6.42.

(3) For the *LCR* circuit in Fig. 6.43, with $V_1(t) = 5 \sin 2\pi \, ft$, $R_1 = 50 \, \Omega$, $C_1 = 10 \, uF$, $L_1 = 100 \, mH$, determine the resonant frequency, loaded Q-factor, and maximum peak current. Use a **PARAM** part/**Parametric Sweep** to vary the capacitance to investigate the effect on the resonant frequency and the Q-factor ($Q = 1/r \sqrt{L/C}$).

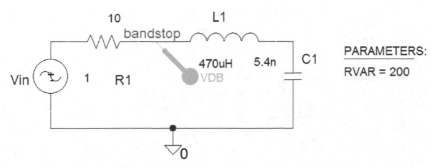

FIGURE 6.43: Series-tuned LCR circuit

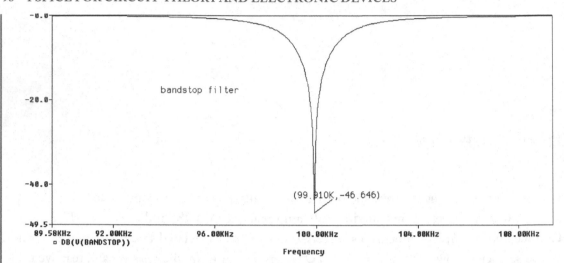

FIGURE 6.44: Bandstop series-tuned circuit

(4) The bandstop series-tuned circuit in Fig. 6.44 has a transfer function given as

$$H(s) = \frac{sL + 1/sC}{R + sL + 1/sC} = \frac{s^2LC + 1}{sCR + s^2LC + 1} = \frac{s^2 + 1/LC}{s^2 + sR/L + 1/LC}$$

$$= \frac{s^2 + \omega_0^2}{s^2 + (\omega_0/Q)s + \omega_0^2} \tag{6.60}$$

A bandpass transfer function is

$$H(s) = \frac{R}{R + sL + 1/sC} = \frac{sCR}{sCR + s^2LC + 1} = \frac{sR/L}{s^2 + sR/L + 1/LC}$$

$$= \frac{s(\omega_0/Q)}{s^2 + (\omega_0/Q)s + \omega_0^2} \tag{6.61}$$

The bandstop filter response as shown in Fig. 6.45.

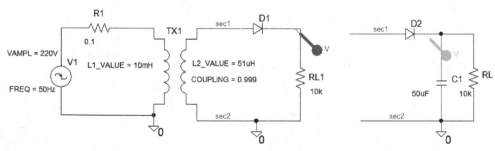

FIGURE 6.45: Bandstop frequency response

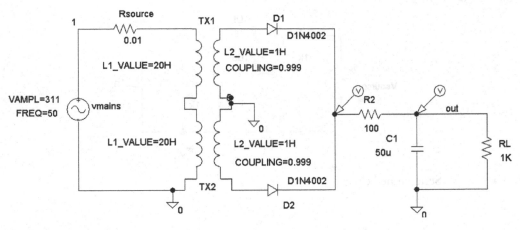

FIGURE 6.46: Rectification

Show that (6.61) is the same as subtracting (6.61) from 1.

(5) In power factor correction, verify the relationship between apparent power (VA), real power (Watts), and reactive power (VARS) as $VA = \sqrt{P^2 + Q^2}$.

(6) Investigate the power supply in Fig. 6.46.

(7) Investigate the center-tap transformer in Fig. 6.47 using **XFRM_LINEAR** parts.

(8) Reduce the circuit in Fig. 6.48 to an equivalent series circuit, by converting the parallel combination of C_1 and R_2. Hence, show that the resonant frequency is $\omega_0 = \sqrt{(\frac{1}{C_1 L_1} - \frac{1}{C_1^2 R_2^2})}$. Calculate the resonant frequency, Q-factor, and the -3 dB Bandwidth for $R_1 = 10\ \Omega$, $R_2 = 100$ kΩ, $L_1 = 1$ mH, $C_1 = 1$nF. Simulate and compare measured results to the calculated values.

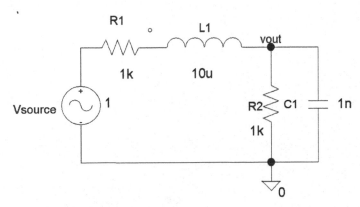

FIGURE 6.47: Centre-tap transformer

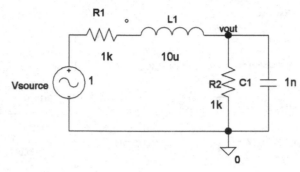

FIGURE 6.48: Series-tuned circuit

CHAPTER 7

Semiconductor Devices
and Characteristics

7.1 SEMICONDUCTOR DEVICES

Electronic devices are investigated by examining the device input and output current–voltage characteristics. Device parameters, such as the input and output resistances, current and voltage gains etc, are measured from the characteristics plotted. PSpice allows nested sweep operations where more than one parameter is required to be swept at any given time. For example, plotting the output transistor characteristic requires sweeping the output voltage for a range of swept input base currents.

7.2 THE FORWARD AND REVERSE-BIASED DIODE CHARACTERISTIC

Fig. 7.1 shows a low-power diode (John Bardeen 1908–1991) in series with a current limiting resistor R_1. The diode current, junction voltage V_d, and temperature T (degrees Kelvin) are related:

$$I_d = I_0(e^{qV_d/kT} - 1) \tag{7.1}$$

The reverse saturation current is $I_0 \cong 10^{-12}$ Amps, q is the charge $= 1.6 \times 10^{-16}$ C, and the Boltzmann constant k is equal to 1.38×10^{-23} J/T.

Select **DC Sweep** and enter the parameters shown in Fig. 7.2. We may plot diode current versus the voltage across the cathode–anode by placing a current marker as shown and sweeping the input voltage.

Press **F11** to simulate to plot the diode characteristic of diode current versus the swept input voltage. This is not the correct characteristic, however as it is necessary to change the swept input voltage to the voltage across the diode **V1(D1)**. Select the space between the x-axis numbers and the menu shown in Fig. 7.3 should appear.

Select **Axis Variable** and highlight **V1(D1)**, which automatically places it in the **Trace Expression** box as shown in Fig. 7.4. Press **Ok**.

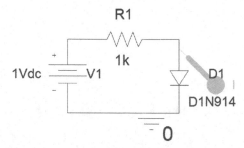

FIGURE 7.1: Circuit for plotting diode characteristic

Fig. 7.3 will appear again so select **Scale Linear** and **press OK**. Place cursors as shown in Fig. 7.5 and measure the DC resistance on the forward-biased diode characteristic.

7.3 DIODE PARAMETERS

The following parameters are obtained from the diode characteristic:

- The DC resistance (static resistance),
- The dynamic resistance, and
- The average AC resistance

The diode DC resistance $V/I = 212 \; \Omega$ is measured from the characteristic shown in Fig. 7.6. To obtain the dynamic ac resistance, use a second cursor on a magnified portion of the characteristic.

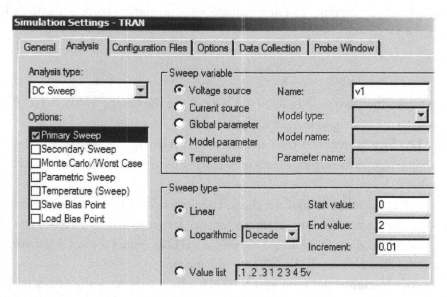

FIGURE 7.2: Sweep parameters

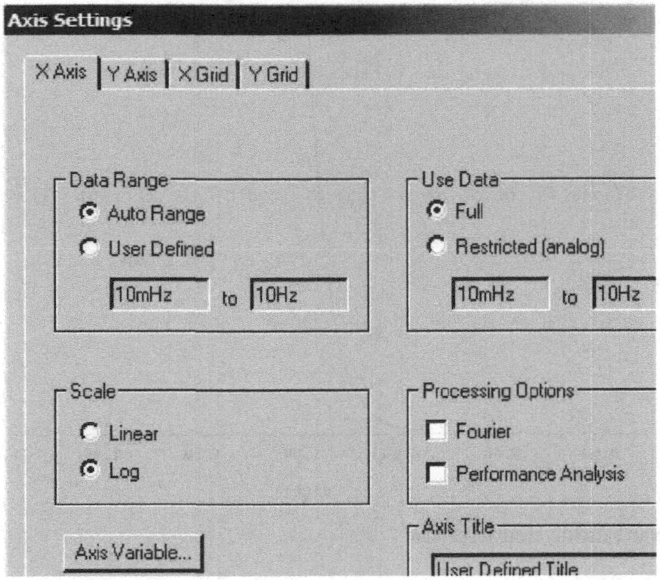

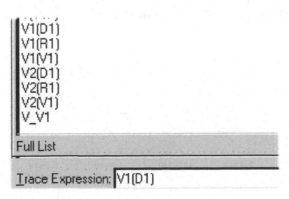

FIGURE 7.3: Changing the *x*-axis variable

An AC voltage across the diode produces a larger resistance than that measured in Fig. 7.6. Increase the separation between the two cursors and measure the average resistance which is a much larger resistance than measured from the inverse of the slope.

7.4   DC LOAD LINE

The load line plotted in Fig. 7.7 is a useful technique for selecting quiescent DC operating conditions.

FIGURE 7.4: Selecting a variable

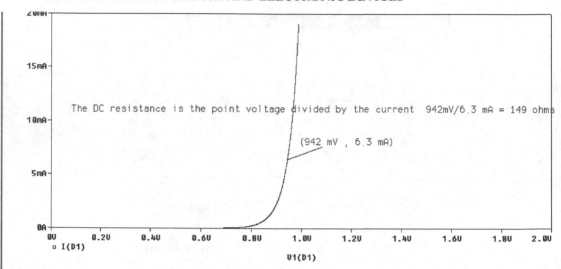

FIGURE 7.5: Forward diode characteristic

Increase the swept input voltage to 15 V and superimpose a load line on the characteristic by selecting **Trace/Add** (a trace menu in Probe) and entering **(5V -V1(D1))/500** in the **Trace Expression**. Read the Q-point values from the intersection of the 1 kΩ load line and the diode characteristic. An ordinary diode is not designed to operate in the reverse breakdown region, but Fig. 7.8 shows the result when the DC is increased past the breakdown region.

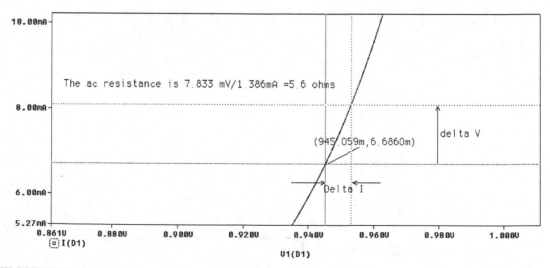

FIGURE 7.6: Magnified section to measure delta changes in V and I

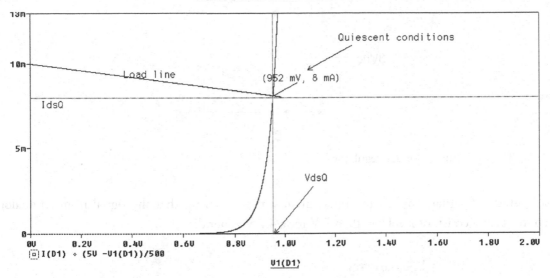

FIGURE 7.7: Determine the quiescent conditions using the load line

7.5   VOLTAGE REGULATION

Zener diodes (William Shockley 1910–1989) are designed to operate in the reverse breakdown region because this region is heavily doped when compared to conventional diodes and so produces a smaller breakdown voltage. The junction voltage remains almost constant even when the reverse diode current changes. This simple regulator may be applied to limiting the

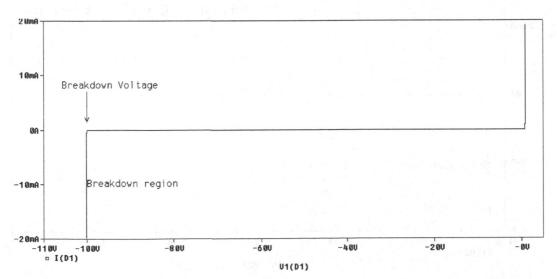

FIGURE 7.8: Reversed-biased diode characteristic

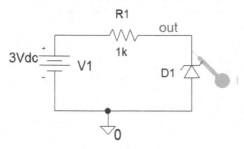

FIGURE 7.9: Simple voltage regulator

amplitude of signals applied to digital circuits, thus ensuring that the signal from an analog source is at a constant level less than 5 V for a 4.7 V zener diode.

7.5.1   Zener Diode Characteristic

The voltage regulator in Fig. 7.9 shows the zener diode operating in its reverse bias mode.

The reverse diode characteristic is obtained by sweeping the voltage V1 in the reverse direction (see Fig.7.2) with the following values: **Analysis type: DC Sweep, Sweep Variable = Voltage source, Name: V1, Linear, Start Value = −40, End Value = 40, Increment = 0.1.** Press **F11**, and in Probe, change the *x*-axis variable from **V_V1** to **V2(D1)**. Fig. 7.10

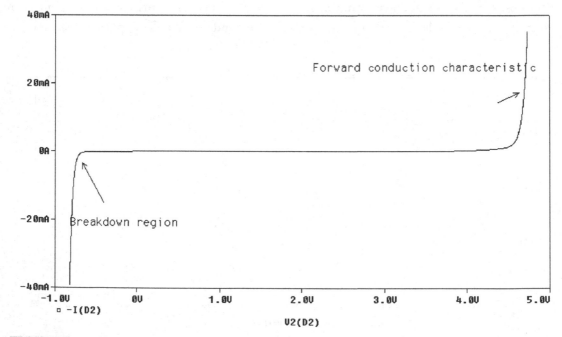

FIGURE 7.10: Zener characteristic

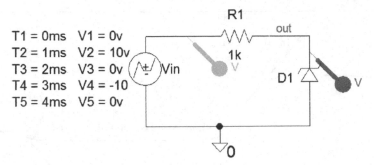

T1 = 0ms   V1 = 0v
T2 = 1ms   V2 = 10v
T3 = 2ms   V3 = 0v
T4 = 3ms   V4 = -10
T5 = 4ms   V5 = 0v

FIGURE 7.11: 4.7 V regulator

shows the complete diode $I-V$ characteristic. V_z is the zener breakdown voltage in the reverse direction. The forward-biased characteristic is similar to a conventional diode. A small zener diode impedance, $Z_s = dV_s/dI_s$, in the breakdown region, makes it useful as a voltage reference (An ideal voltage source resistance is zero).

The device maintains a constant voltage across its terminals even though the current through it is changing. However, the reverse current must be sufficiently large to keep it operating in the breakdown region, thus ensuring correct circuit operation. The minimum zener current $I_z(\min)$ is approximately 0.1 (P_z/V_z). If the diode keeps working, then the diode power dissipation $P_z = V_z^* I_z$ must not be exceeded. The unregulated input voltage must be large enough to supply sufficient diode breakdown current. The load R_L across the regulator cannot be too small because it would draw excessive current, leaving insufficient current to keep the diode in the breakdown region.

7.5.2   Zener Diode Regulation

Fig. 7.11 is a voltage piece-wise linear (VPWL) part where the input signal source, after selection and **Rclicking/Edit Properties**, has the time–voltage pair values entered in the spreadsheet.

Fig. 7.12 shows how the zener diode limits the amplitude of the output signal.

To plot the output voltage for a range of input voltage–the transfer function) shown in Fig. 7.13, we need to change the x-axis from *Time* to the input voltage v1(Vin).

We do this from Probe by selecting an x-axis number and from the menu select **Axis Variable**. Highlighting **V1(Vin)** from the list automatically places it in the **Trace Expression** box, so press **OK**.

7.6   SILICON-CONTROLLED RECTIFIER

A thyristor is a silicon-controlled rectifier (SCR) that can be used to control power to devices such light bulbs, drills etc. We need to obtain the I-V characteristics in order to understand

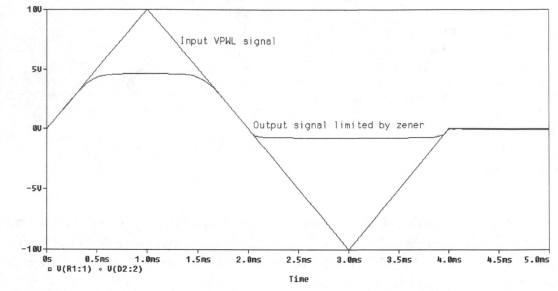

FIGURE 7.12: Input and output signals

the device operation. Draw the SCR schematic and set the Primary and secondary sweeps as shown in Fig. 7.14.

Press the simulation **F11** key to plot the Press **F11** to plot the SCR characteristic as shown in Fig. 7.15. To investigate for different swept ranges, use the log command in Probe to separate the key strokes automatically. Add a new y-axis to accommodate the gate current.

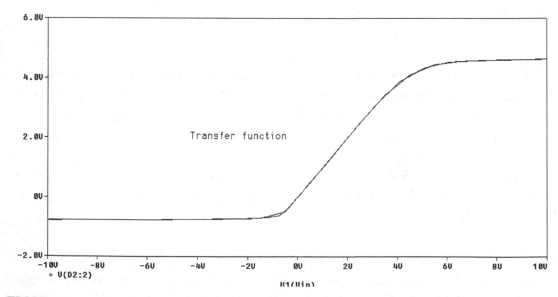

FIGURE 7.13: Transfer function

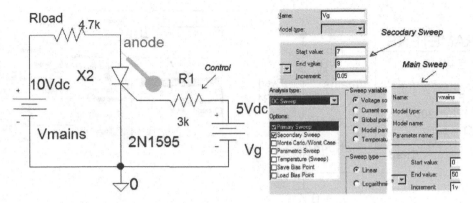

FIGURE 7.14: SCR schematic for plotting the I-V characteristics

7.7 TRIAC CONTROLLER

Draw the Triac schematic in Fig. 7.16 and set the **Parametric Sweep** parameters as shown. The variable resistance, Rvar, controls the power in the load, Rload. We see in Fig. 7.17 how the power is halved when the resistance is increased to 9 kohms.

7.8 THE BIPOLAR TRANSISTOR

Three-terminal transistor devices (Walter Brattain 1902–1987), such as the bipolar junction transistor (BJT) or the field effect transistor (FET), are characterized by examining input and output characteristics. These characteristics are plotted by sweeping external DC supplies using

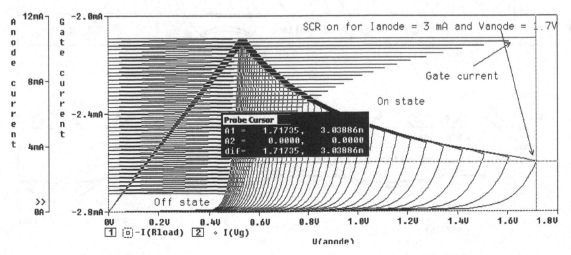

FIGURE 7.15: Triac signals

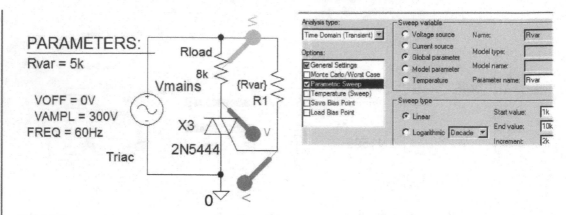

FIGURE 7.16: Triac controller schematic

nested DC sweeps (allows more than one parameter to be swept at any given time). Parameters measured from the characteristic are then used to design amplifiers with specific gains, input and output impedances, quiescent bias points, etc.

7.8.1   The Input and Output BJT Characteristics

A transistor may be connected in three modes: common-emitter, common-base, and common-collector. Fig. 7.18 shows a common-emitter connected transistor, where the emitter is "common" to the input and output circuits. A current source connected to the base

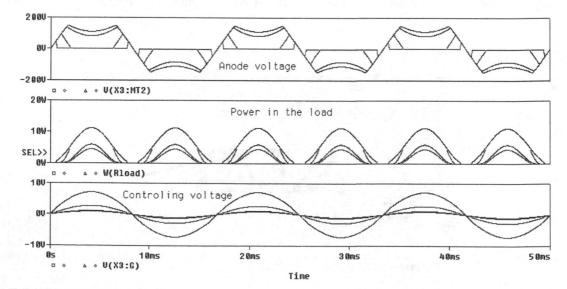

FIGURE 7.17: SCR waveforms

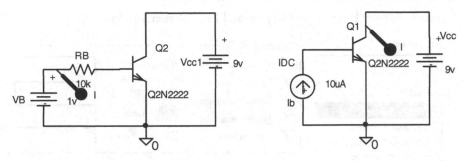

FIGURE 7.18: Common emitter mode

terminal is simulated as a DC voltage source VDC part in series with a high resistance *RB* (Or we may also use an **IDC** current source part as shown). To obtain the input characteristic, sweep the input voltage source by selecting the **Analysis Setup** menu and ticking **DC Sweep**.

The input characteristic is a plot of base-emitter voltage versus base current shown in Fig. 7.19. However, be careful and plot the emitter-base voltage and not the swept input voltage. After simulation, change the *x*-axis by selecting the *x*-axis variable, and in the **Trace Expression** box, change **V_VB** to **V(RB:2)**. Similarly, from the **Trace** menu, add the *y* variable **I(RB)** in the **Trace Expression** box. The transistor ac input impedance h_{ie} is obtained from the inverse of the slope of the characteristic over a small region.

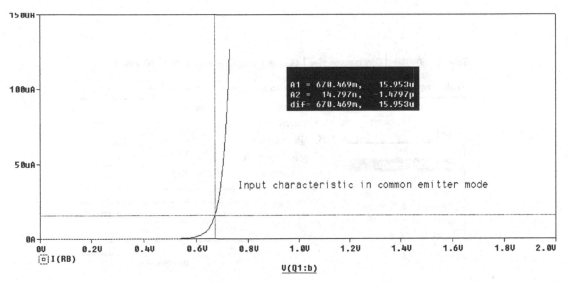

FIGURE 7.19: The input characteristic

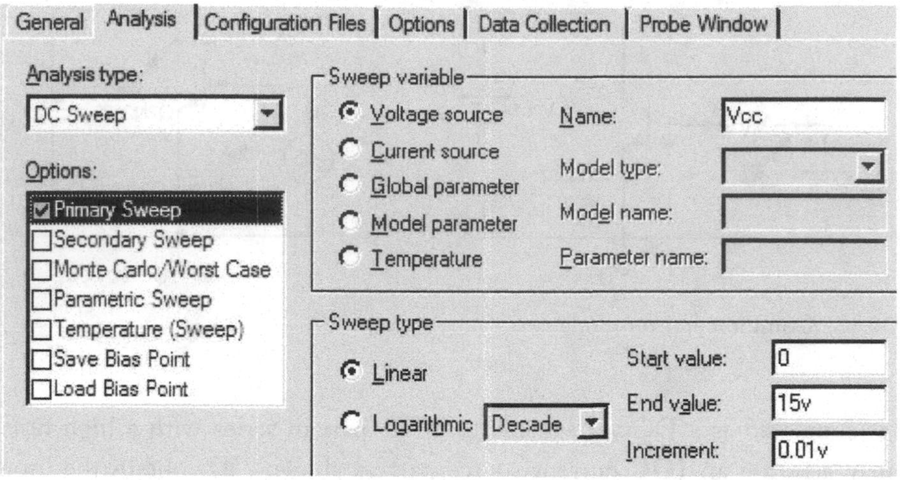

FIGURE 7.20: Enter the Primary base current sweep parameters

7.8.2 The Output Characteristic

The output characteristic is obtained by sweeping the collector-emitter voltage **VCC**. The input current source is swept over a range of base currents using a secondary nested sweep for a range of swept collector voltages. From the **Analysis** menu, select **DC Sweep** shown in Fig. 7.20 and set the **Vcc** parameters in the **Primary Sweep** menu.

Select the **Secondary Sweep** and enter the values shown in Fig. 7.21.

Press the **F11** button to obtain the output characteristic as in Fig. 7.22.

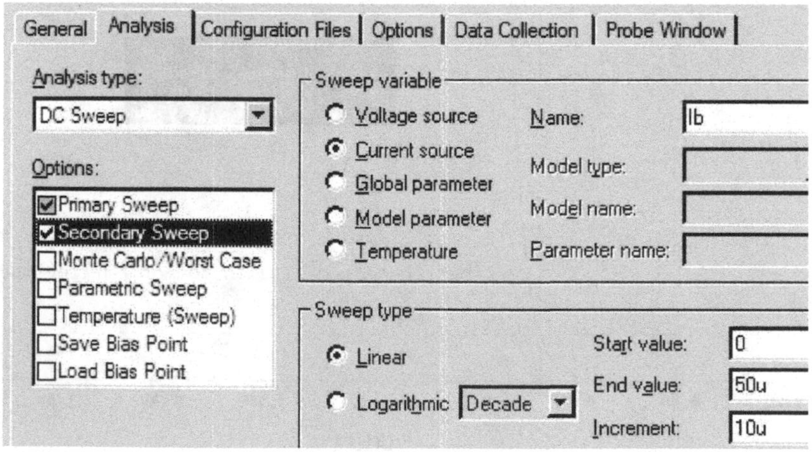

FIGURE 7.21: Setting the collector sweep voltage parameters

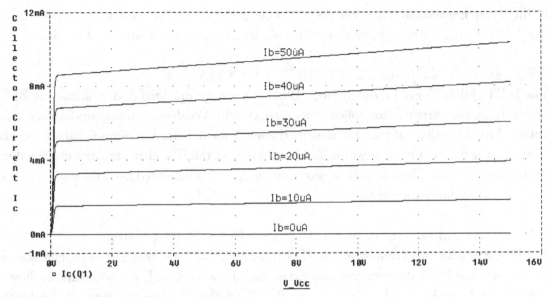

FIGURE 7.22: Output characteristic in CE mode

7.8.3 DC Load Lines

Load lines are useful for selecting quiescent operating conditions. Obtain the output characteristic as before but increase the collector voltage, Vcc, to 15 V. To superimpose a load line on the output characteristic as shown in Fig. 7.23, click the **Probe Trace** menu and select **Add/Trace**.

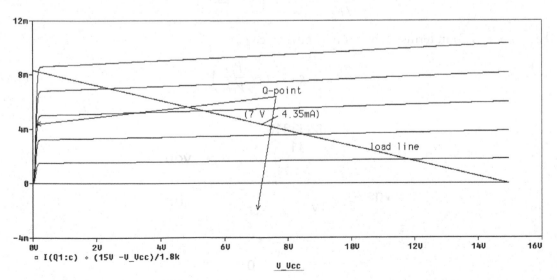

FIGURE 7.23: Transistor load line

In the **Trace Expression** box, enter **(15V -V_Vcc)/1.8k** to superimpose a 1.8 kΩ load line on the characteristic. Select a Q-point to permit maximum signal swing without distortion.

7.9   JUNCTION FIELD-EFFECT TRANSISTOR

The JFET (Julius Edgar Lilienfeld 1881–1963) is a voltage-operated device, unlike the BJT, where the output current is controlled by an input current. We define a trans-conductance, g_m, obtained from the slope of the FET transfer (input) characteristic. The output characteristic is plotted using a nested loop, as was utilized for obtaining the BJT characteristics explained in the previous section. The output transistor resistance, r_{DS}, is measured from the inverse of the slope of the output characteristic.

7.9.1   The Common Source JFET Transistor Input Characteristic

The J2N3819 field-effect transistor is an N-channel device with a P-type gate so the gate source is reversed biased with the battery connected in the common source JFET schematic as shown in Fig. 7.24. Note: Reverse both batteries for the J2N3820 FET, since this is a P-channel with an N-type gate.

The drain current, gate–source voltage, and pinch-off voltage are related as

$$I_{ds} = I_{dss}\left(1 - \frac{V_{gs}}{V_{PO}}\right)^2 \tag{7.2}$$

Differentiate the drain current (7.2) with respect to the gate–source voltage to yield:

$$\frac{dI_{ds}}{dV_{gs}} = \frac{2I_{dss}}{-V_{po}}\left(1 - \frac{V_{gs}}{V_{PO}}\right) \tag{7.3}$$

Expressing (7.3) in terms of the trans-conductance g_m:

$$g_m = g_{mo}\left(1 - \frac{V_{gs}}{V_{PO}}\right) \tag{7.4}$$

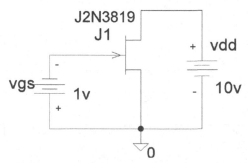

FIGURE 7.24: Circuit for obtaining the FET characteristics

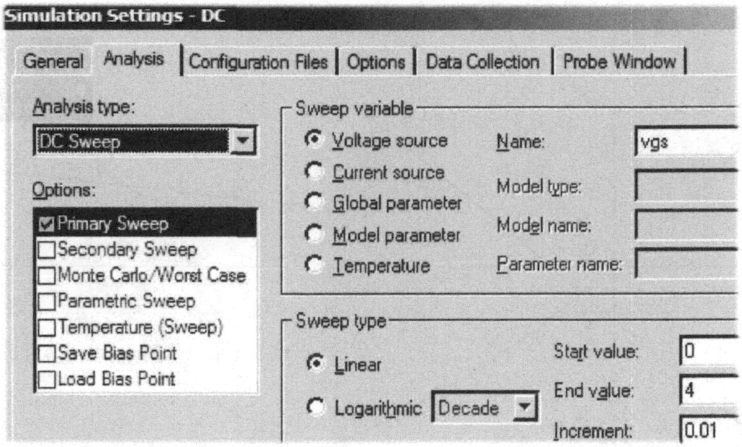

FIGURE 7.25: Setting the parameters for DC sweep

Thus the mutual trans-conductance is a function of the gate–source voltage. Plot the rate of change of drain current with the gate–source voltage V_{gs}. The trans-conductance g_m is obtained from the slope of the transfer characteristic *around the selected quiescent operating point*.

$$g_m = \left. \frac{\Delta I_{DS}}{\Delta V_{GS}} \right|_{V_{DS}=10\text{V}} \tag{7.5}$$

Place a current marker at the drain. From the **Analysis Setup** menu in Fig. 7.25, select **DC Sweep** and tick Voltage source. Enter v_{GS} in the **Name** box and set the parameters as shown.

Press **F11 to simulate**. Measure the DC current and voltage conditions after simulating by using the voltage **V** and current **I** icons. In Fig 7.26, select a quiescent drain–source current Idsq approximately half way between 0 and the maximum drain current I_{dss}. This is made easier if we add a load line as explained in the next section.

7.9.2   Adding a Load Line to the Transfer Characteristic

The common source JFET amplifier is self-biased by the voltage developed across a resistance placed from source to ground. The biasing voltage in previous circuits is not now required since the current flowing through the resistance produces a quiescent bias voltage from the voltage drop across R_1. It is called self-biased because it uses its own source current to set up a bias voltage across the resistance. From the input characteristic, select a quiescent current approximately half way up the y-axis and drop a perpendicular to the x-axis and read the voltage Vdsq. From these values we determine a suitable value for the source–gate biasing resistance R_1.

$$V_1 = -V_{GSQ} = -I_{DSQ}R_1 \Rightarrow R_1 = V_{GSQ}/I_{DSQ} = 0.97\text{V}/5.4\text{ mA} = 180\ \Omega \tag{7.6}$$

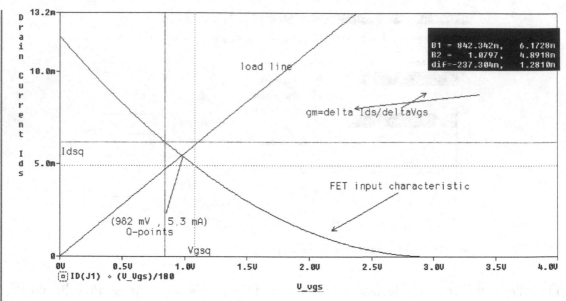

FIGURE 7.26: Transfer function with load line

We can superimpose a biasing 180 Ω load line on the mutual characteristic as shown in Fig. 7.26, by selecting the **Trace** menu in Probe and **Add Trace**. In the **Trace expression** box, add the expression **(V_Vgs)/180**. Reset the current axis to 12 mA by clicking on any y-axis variable or the space beside the variable and select **User Defined** and place the new y-axis limits. In practice, the DC voltage source is replaced by a large resistance and not with a short circuit, as this would short out any ac input signal connected across the gate–source terminals. *The value of RG in Fig. 7.27 is 1meg* (with no space between 1 and MEG), *and not 1m, which is a milliohm.* No DC voltage is dropped across this resistance since the current in the input circuit is almost zero (= 1.249 picoamp ≈ 0) and will not change the DC conditions.

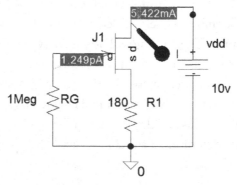

FIGURE 7.27: Inserting a 1 MΩ resistance

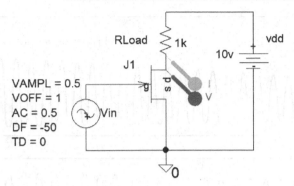

FIGURE 7.28: VSIN generator

Measure the quiescent conditions obtained by the source resistance R_1, and compare V_{GSQ} and I_{DSQ} to those selected using the load line. Note: Adding a load resistance RL in the drain circuit will change the quiescent conditions by reducing the DC current and hence the bias across R_1, so we must factor this into the final design.

7.9.3   Quiescent DC Operating Point

A JFET amplifier is normally biased in a linear region of the characteristic, so a good choice for the quiescent DC point is in the constant current region. If the selected Q-point is too close to the pinch-off voltage, the signal will clip at certain input signal levels. A good choice of quiescent biasing minimizes signal distortion and allows for a symmetrical signal swing. The JFET is biased so that the drain–source voltage is roughly halfway between the supply voltage and the pinch-off voltage. To demonstrate incorrect bias, connect a **VSIN** generator as shown in Fig. 7.28 where the bias is set by the generator offset value. The generator produces a sine wave whose amplitude increases exponentially with time and whose instantaneous output is defined as

$$v(t) = VAMP\sin(2\pi(FREQt + PHASE/360))e^{-tDF} + VOFF \qquad (7.7)$$

In (7.7), the variable generator parameters are capitalized. **VOFF**set provides a DC bias and an exponentially increasing sine wave with time is achieved by changing **−DF** to **DF**.

The result of incorrect biasing in a JFET amplifier is shown in Fig. 7.29. Clipping occurs when the sine signal increases in amplitude to a certain level.

7.9.4   JFET Output Characteristic

DC nested sweeps contain two nested loops to plot the JFET output characteristics. The second sweep variable is selected once the primary sweep value has been specified in the **DC Sweep**

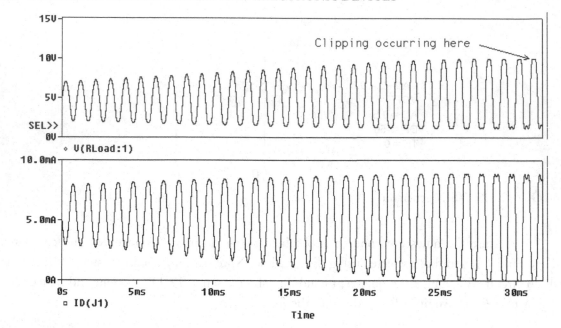

FIGURE 7.29: Output current and voltage waveforms

Analysis type box. Set up the circuit shown in Fig. 7.30 and place a current marker at the drain.

Set up a nested sweep from the **Analysis type** menu by selecting **DC Sweep**. Tick **Voltage source** and enter the **vdd** parameters for the main loop as shown in Fig. 7.31.

Tick **Secondary Sweep** as shown in Fig. 7.32 and set the **Sweep variable** parameters for the gate–source inner sweep voltage Vgs.

Simulate with the **F11** key. To superimpose a load line on the family of output characteristics, select the **Trace** menu, and add a trace with **Add Trace**. The x-axis variable in Probe

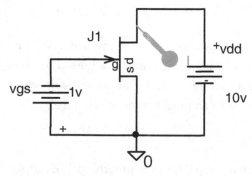

FIGURE 7.30: Output characteristic circuit

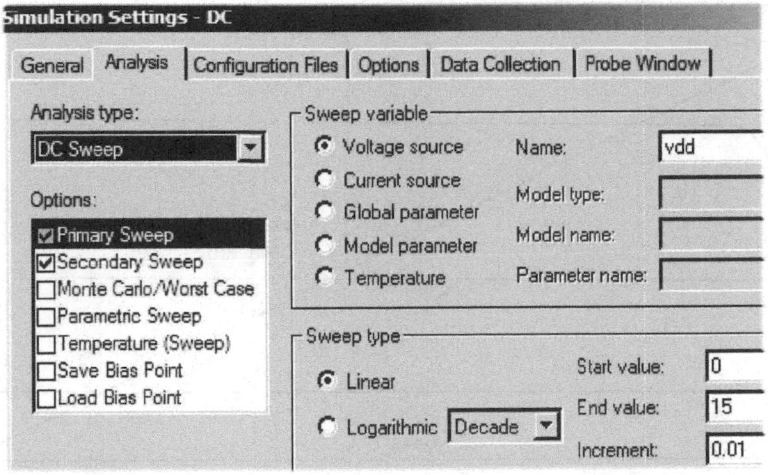

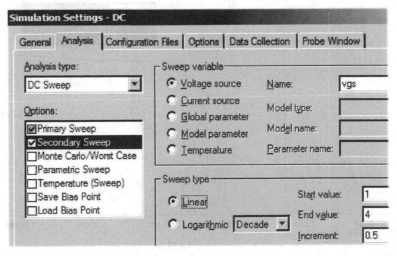

FIGURE 7.31: Setting up the outer loop

is Vdd (0–15) V, so in the **Trace Expression** box, enter **(15V -V_Vdd)/2.2k** to superimpose a 2.2 kΩ load line as shown in Fig. 7.33.

The JFET output resistance is measured from the inverse of the slope of the output characteristic defined as

$$r_{DS} = \frac{\Delta V_{DS}}{\Delta I_{DS}}\bigg|_{V_{GS}=V_{GSQ}}$$

Enlarge a section of the characteristic as shown in Fig. 7.34 by selecting the y-scale or use the magnifying tool.

FIGURE 7.32: The nested sweep parameters

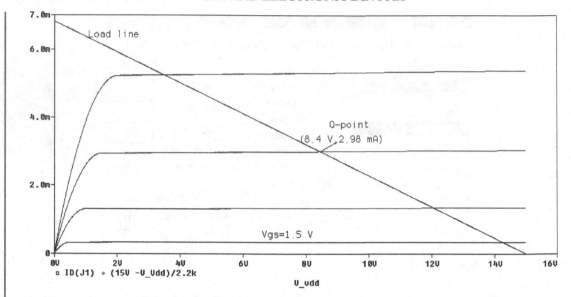

FIGURE 7.33: JFET output characteristic for a 1 kΩ load line

7.9.5   Effect of Temperature on the JFET Transfer Characteristic

Semiconductor devices behave differently over a range of temperature (Ludwig Boltzmann 1844–1906) values, so it is important to investigate circuit behavior for a temperature sweep using the schematic in Fig. 7.35.

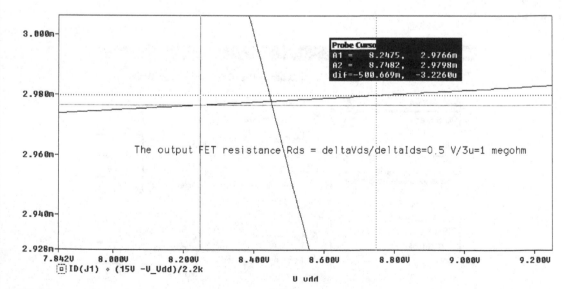

FIGURE 7.34: Enlarged section of the output characteristic

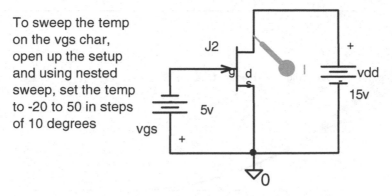

To sweep the temp on the vgs char, open up the setup and using nested sweep, set the temp to -20 to 50 in steps of 10 degrees

FIGURE 7.35: Investigating device change with temperature

From the **Analysis tab** menu, select **DC Sweep** and tick **Temperature (Sweep)** with parameters as in Fig. 7.36.

Press **OK,** and the effects of temperature on the characteristic are displayed in Fig. 7.37.

Differentiating the drain current $I_{ds} = I_{dss}[1 - (V_{gs}/V_{PO})^2]$, yields $\Rightarrow \frac{dI_{ds}}{dV_{gs}} = \frac{2I_{dss}}{-V_{PO}}(1 - \frac{V_{gs}}{V_{PO}}) = g_m = g_{mo}(1 - \frac{V_{gs}}{V_{PO}})$. The mutual conductance is a function of the gate source voltage V_{gs} and is measured from the transfer characteristic. In Probe, press the **"Add a trace"** icon and the screen in Fig. 7.38 appears. Select the differential operator **D( )** from the list on the right, and in the brackets place the selected variable **ID(J1)** from the list on the left.

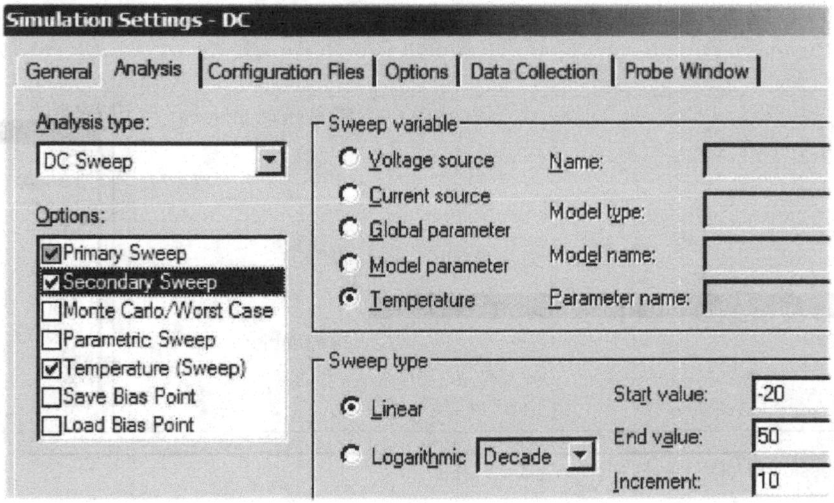

FIGURE 7.36: Select temperature as the DC nested sweep

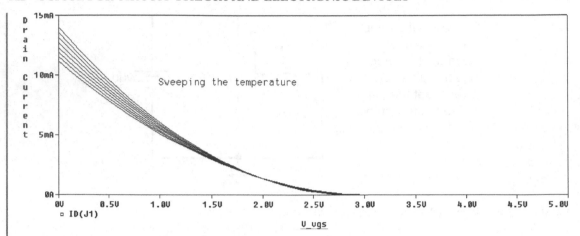

FIGURE 7.37: The effect of temperature sweep on the FET transfer curve

7.10   THE D OPERATOR

The D operator is a useful function available from Probe because it allows you to plot the differential of a function. Fig. 7.39 is a plot of the differential drain current showing how g_m increases linearly with gate–source voltage.

7.11   EXERCISES

(1) Plot the inverse of the differential of the diode current/diode voltage. Use the differential operator **D** in Probe as **1/(D(I(D1)/V1(D1)))** to plot the inverse of the diode slope characteristic.

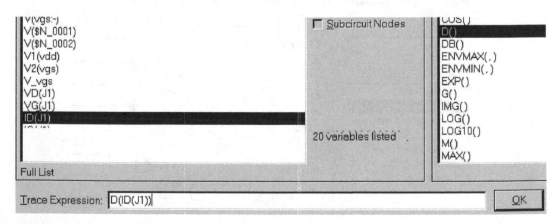

FIGURE 7.38: Select the differential of the drain current

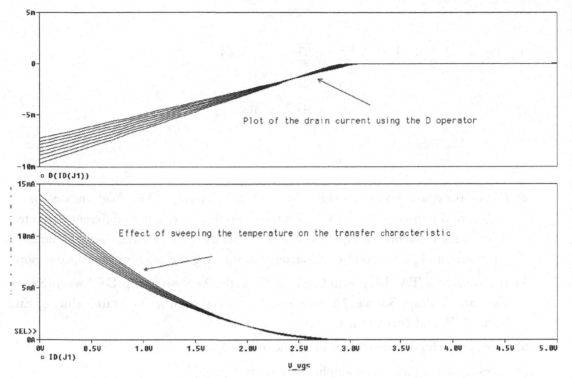

FIGURE 7.39: Differential of drain current

(2) In Fig. 7.40, investigate the diode turn on/off times using a **VPULSE** generator with a very fast ±5 V pulse (I suggest **TR = TF** = 0.1ns, **PW** = 10 ns, **PER** = 20 ns). Plot the diode current.

(3) Plot the Zener diode impedance in the different regions of the characteristic. Use the **D** operator.

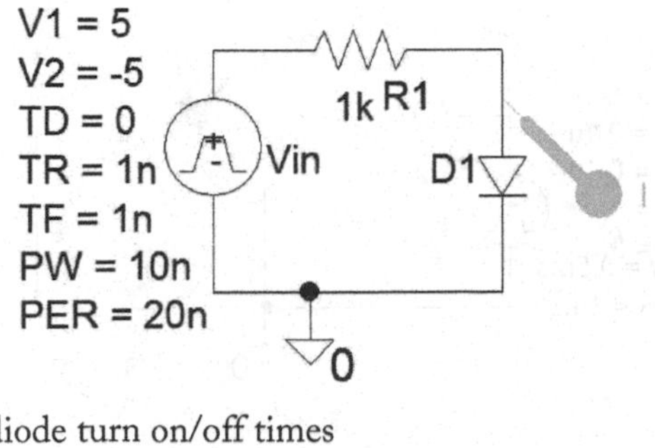

V1 = 5
V2 = -5
TD = 0
TR = 1n
TF = 1n
PW = 10n
PER = 20n

Vin

1k R1

D1

0

FIGURE 7.40: Measuring diode turn on/off times

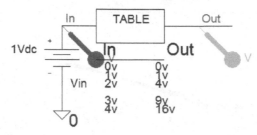

FIGURE 7.41: The Table part

(4) Obtain the characteristics for the SCR and TRIAC devices. Download a model for the unijunction transistor (UJT). This is a three-terminal device with different characteristics to the conventional bipolar transistor. Construct a UJT circuit symbol, junction schematic, and plot the output characteristics showing the negative resistance region.

(5) Investigate the **TABLE** part in Fig. 7.41. From the **Analysis Setup/DC Sweep/Sweep Variable /Voltage Source, Name** = vin. The sweep parameters: **Start Value: 0, End Value** 10V, and **Increment 1**.

(6) Investigate the BJT switching schematic in Fig. 7.42.

(7) Investigate the differential amplifier shown in Fig. 7.43.

(8) Investigate root locus plotting using **DIFF** and **Laplace** parts in Fig. 7.44. Analysis tab to **Analysis type: AC Sweep/Noise, AC Sweep Type** to **Logarithmic, Start Frequency** = 0.001, **End Frequency** = 10, **Points/Decade** = 10,001.

Press **F11** to produce the locus plot shown in Fig. 7.45 but change the x-axis to **R**(v(out)) and the y-axis to **IMG**(v(out)). Both of these functions are available from **Trace/Add Trace** in Probe. However, in the latest edition of PSpice, there is now

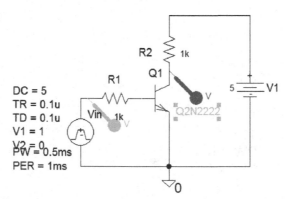

FIGURE 7.42: BJT switch

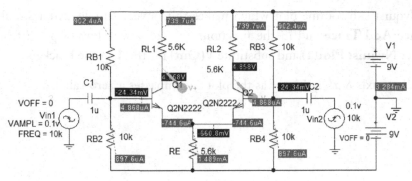

FIGURE 7.43: Differential amplifier

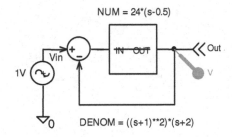

FIGURE 7.44: A closed-loop control system

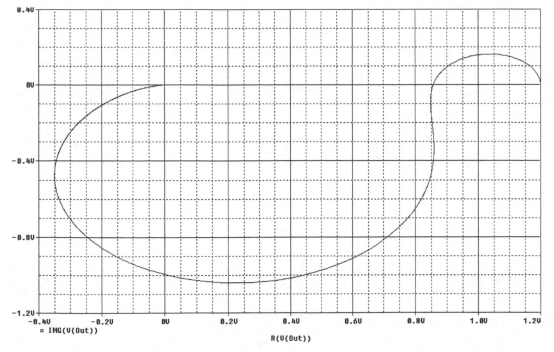

FIGURE 7.45: Change y-axis to IMG(v(out)) and x-axis to R(v(out))

a Nyquist sub routine plot which makes things very simple. After simulation, select **Trace/Add Trace** and in the functions or macros window (top right hand corner), select **Nyquist Plot(1)** and substitute **V(Out)** for the **1** in the brackets.

Make sure the *x*-axis *Scale* is *Linear* as the plot encompasses a zero value.

CHAPTER 8

Operational Amplifier Characteristics

8.1   IDEAL OPERATIONAL AMPLIFIERS

The normal operational amplifier (opamp) has two-input terminals and a single-output for amplifying the voltage difference between the inverting and noninverting input terminals. An ideal opamp has an infinite input impedance $R_i = \infty \Omega$, zero output impedance $R_o = 0\ \Omega$, and infinite gain, A. A nonideal op-amp has a very high input resistance, very low output resistance, and a very high gain.

8.1.1   The Inverting Operational Amplifier and Virtual Earth

Fig. 8.1 shows a voltage-controlled voltage source (VCVS) **E** part (**Analog.olb** library) modeling the operational amplifier in the linear range. The input impedance is set to 2 MΩ, with an 75 Ω output impedance but with no frequency limitations (see exercise 10 at the end of the chapter). Apply a **1** VDC part and press **F11** to simulate and display the bias conditions. A very large gain 10^{10} produces a voltage at the junction of R_1 and R_2 of 477 pV, which is almost

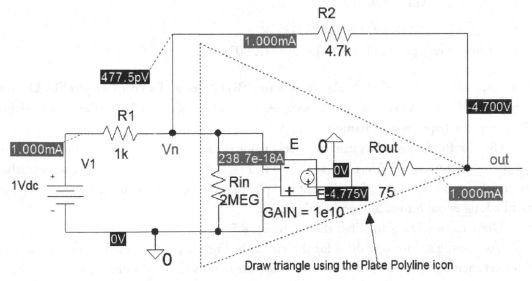

FIGURE 8.1: An ABM op amplifier model

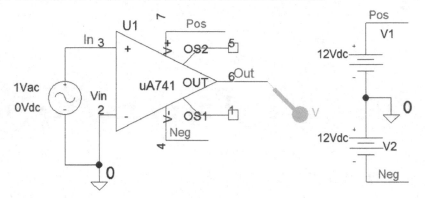

FIGURE 8.2: Open circuit opamp

zero. Hence, this point is called a virtual earth. This simple linear opamp model overcomes the limitations of the evaluation version in a situation requiring a large number of operational amplifiers (there is also an ideal **opamp** part available).

Draw the operational amplifier shown in Fig. 8.2.

The full gain of the ua741 operational amplifier is available in this comparator so that, because of the large gain, a small input voltage will easily saturate the output to the rail voltage. The transfer function characteristics may be read from the output file accessed from the **PSpice/View Output File** menu, or from **Probe/View.**

Small signal characteristics

- $V(Out)/V\_vin = 1.963E+05$
- Input resistance at $V\_vin = 9.963E+05$
- Output resistance AT $V(Out) = 1.494E+02$

However, you must first select the **Analysis Setup/Bias Point** and set the **Output File Options** as shown in Fig. 8.3. **Out** is the output wire segment name using the **Net Alias** icon explained below, and the **Input source name** to Vin.

Use the **Place Net Alias** icon to place a name on a wire. When you open up this menu, type the name **out** in the **Alias** box as shown in Fig. 8.4. This produces a little square which must be dragged over to the wire segment and placed accurately on the wire. The wire segment must be long enough to accommodate the name.

Draw the inverting amplifier shown in Fig. 8.5.

To investigate noise analysis for the circuit in Fig. 8.5, set the parameters as shown in the insert menu and run an AC analysis. To list the main noise contributors, open the output file from Probe.

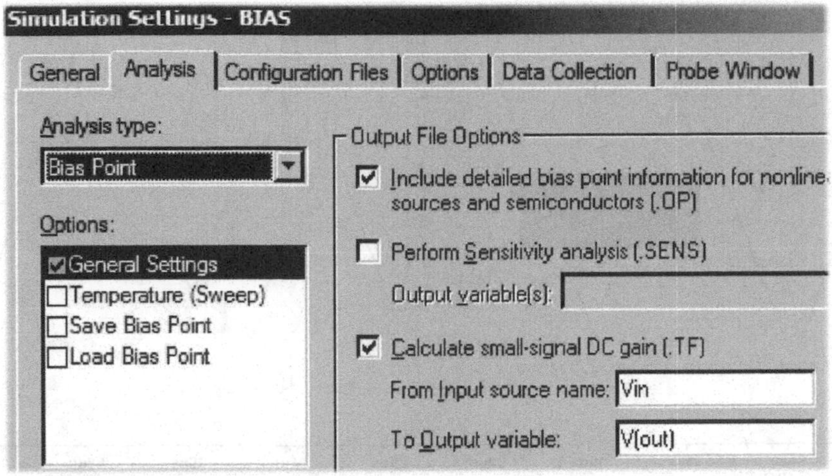

FIGURE 8.3: Transfer function characteristics

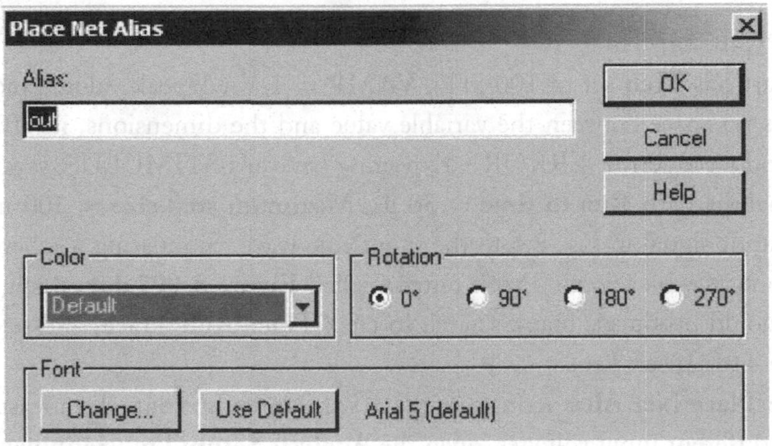

FIGURE 8.4: Placing a wire alias name

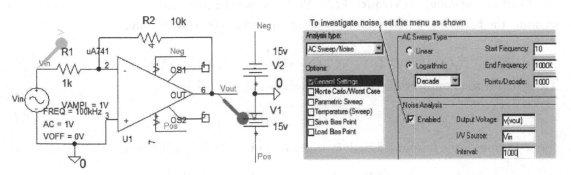

FIGURE 8.5: Inverting mode opamp

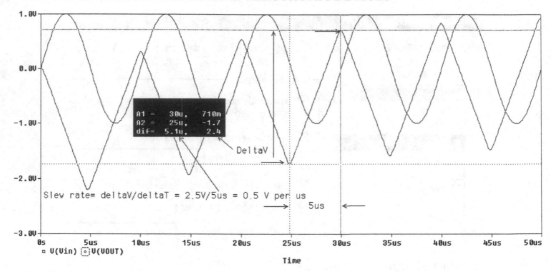

FIGURE 8.6: Slew rate limiting

8.1.2   Slew Rate Limiting

The **VSIN** part has **Freq** set to 100 kHz, **VAMP** = 1 V (A peak value), and **VOFF** = 0. [Note there is no space between the variable value and the dimensions, i.e. 100k, not 100 k as this will produce an error "ERROR – Expecting keyword STIMULUS, saw k"]. Carry out a transient analysis with **Run to time** = 50 us, **Maximum step size** = 100 n and compare the slewed output signal in Fig. 8.6 to the input sine wave. Simulating a schematic produces several files, one of which is the Probe output called **Figure 8-005.dat** which when selected from Probe should produce a blank screen, so click **Trace/Add Trace**, and select the output signal (Or press the Insert key).

Use the **Place Net Alias** icon to name a wire segment **vout**. To measure the output distortion for different input voltages, open the **Analysis Setup/Transient** menu. Place a tick on **Perform Fourier Analysis, Center frequency** = 100 k, **Number of Harmonics** = 10, and **Output Variables:** = **v(vout)**. From Probe, select the icon from the left-hand toolbar to examine the harmonic analysis information at the end of the **.out** text file.

8.2   THE NONINVERTING OPERATIONAL AMPLIFIER

The gain for the noninverting configuration shown in Fig. 8.7 is

$$A_{vc} = V_{out}/V_{in} \qquad (8.1)$$

β is a portion of the output fed back from the output to the input. The output voltage is

$$V_{out} = A(V_{in} - \beta V_{out}) = AV_{in} - \beta AV_{out} \Rightarrow V_{out}(1 + \beta A) = AV_{in} \qquad (8.2)$$

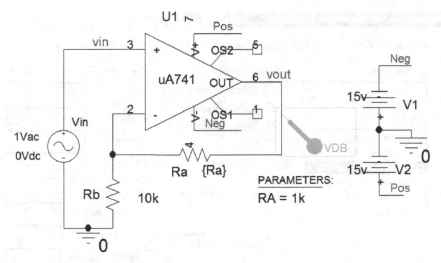

FIGURE 8.7: Noninverting configuration

$$A_v = \frac{V_{\text{out}}}{V_{\text{in}}} = \frac{A}{1 + \beta A} \qquad (8.3)$$

If the open–loop gain is large then we may write (8.3) as

$$A_v = \frac{1}{1/A + \beta} \approx \frac{1}{0 + \beta} \approx \frac{1}{\beta} \qquad (8.4)$$

The feedback factor is

$$\beta = \frac{Rb}{Rb + Ra} \Rightarrow A_v \simeq \frac{1}{\beta} \simeq \frac{Rb + Ra}{Rb} = 1 + \frac{Ra}{Rb} \qquad (8.5)$$

8.2.1 Gain–Bandwidth Product

A **PARAM** part investigates the concept of **GAIN–BANDWIDTH** product and shows how the gain and bandwidth parameters are interlinked i.e. if the gain is reduced then the bandwidth increased and vice versa.

Rclick the **PARAM** part, select **Edit Properties** and add a new row with **Ra** entered in the first column and 1k in the second column. Go back to the schematic and select the resistance Ra and enter {Ra} rather than an actual resistance value. We wish to carry out a frequency response for the different values of the resistor to be swept, so select the **Analysis tab** from the simulation icon and select **Analysis type: AC Sweep/Noise, AC Sweep Type to Linear, Start Frequency = 10, End Frequency = 1000k, Points/Decade = 1000**. Tick **Parametric Sweep** in the **Options** menu and set the parameters shown in Fig. 8.8. Enter the values for Ra in the **Value List** with each value separated by a space.

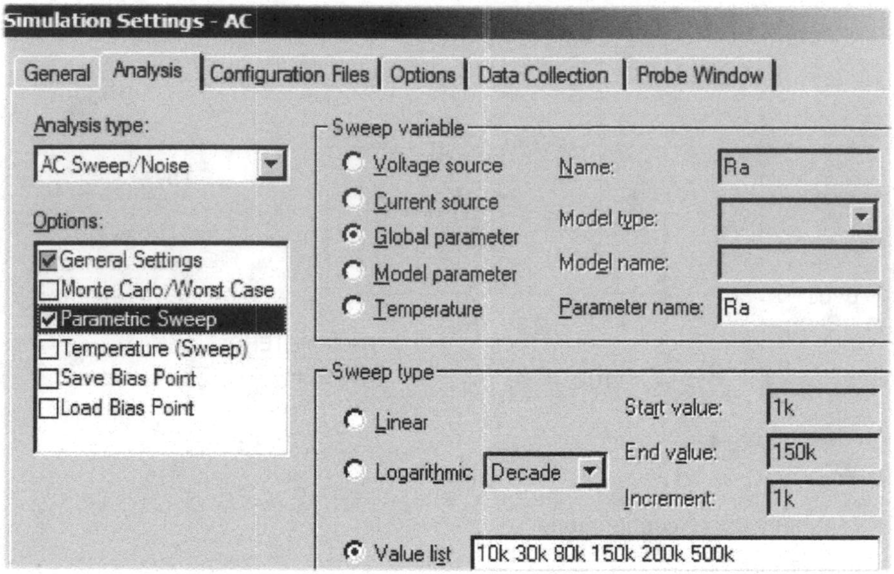

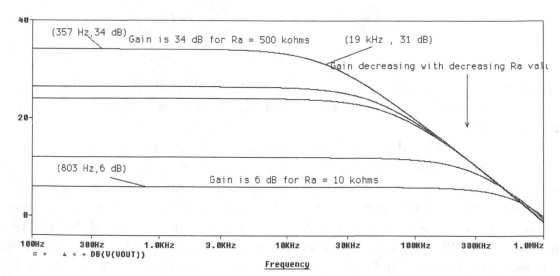

FIGURE 8.8: Using Param and list

Press **F11**, and a Probe screen message "**Available Sections**" appears. Select **OK** to plot the frequency response as shown in Fig. 8.9. Measure the passband gain and −3 dB point on each response.

The gain–bandwidth product is investigated using **Performance Analysis** (PA) available in **Probe/Trace menu**. PA allows you see how a characteristic, such as the gain bandwidth

FIGURE 8.9: Noninverting frequency response

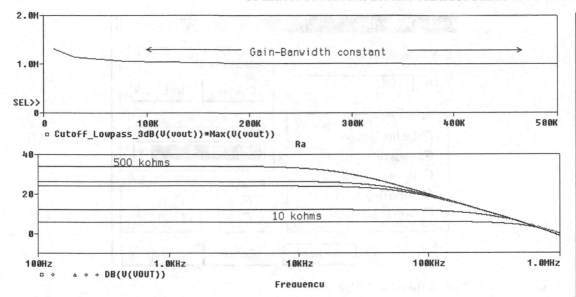

FIGURE 8.10: GB product using Performance Analysis

product in this case, varies when the feedback resistor, Ra is varied. Simulate the previous schematic, and from the Probe output, press **ok** from **Available Sections**. Press the **PA** icon (Two crossed flags) and a further menu headed **Performance Analysis** appears, so press **OK**. Select the **Trace Add** icon and enter **Cutoff_Lowpass_3dB(V(vout))*Max(V(vout))** in the **Trace Expression** box to produce the display in Fig. 8.10.

8.3   AUDIO POWER AMPLIFIERS

NPN and PNP power transistor symbols are created using the **QbreakN** and **QbreakP** parts from the **break** library. Place **QbreakN** and **QbreakP** parts on a new schematic area. Select one of these symbols and it should turn green. **Rclick,** and select **Edit Properties** from the list. From this properties menu, select **QbreakN** in the second column and replace it with TIP31. Select the Display icon to get the **Design Properties** menu where the name is entered as shown in Fig. 8.11. Tick **Name and Value** to display the name and value on the schematic. Press **OK** Select the transistor and **Rclick**. From the displayed list, select **Edit Pspice Model** to run the **PSpice Model Editor**. The demo version will not allow you to save the new power transistor however, but if you are using the full version then you can save it to **yourcircuitname.lib**. Save the results by selecting **Part** and **Save to Library**. The contents of the .**lib** file are shown in Fig. 8.12.

The first line is the model name. However, we can still use the power transistor from **MYLIB.lib** that was added previously. Repeat this procedure for the TIP32 power transistor

Display Properties

Name: Implementation

Value: TIP31|

Font
Arial 5 (default)

[Change...] [Use Default]

Display Format
- ○ Do Not Display
- ● Value Only
- ○ Name and Value
- ○ Name Only
- ○ Both if Value Exists

Color
Default ▼

Rotation
- ● 0°   ○ 180°
- ○ 90°   ○ 270°

[OK] [Cancel] [Help]

FIGURE 8.11: Type in the name TIP32

and draw the class-B complementary symmetry push–pull amplifier schematic in Fig. 8.13. Q_1 conducts for positive input signals and Q_2 conducts for negative input signals. The **VSIN** part has a peak value of 7 V and **FREQ** = 1k.

Place current markers to plot the current drawn from each power supply by the transistors. The average power is plotted in Fig. 8.14 using the **AVG()** operator from the **Trace/Add Trace** menu in Probe as **AVG(I(Rload)*V1(Rload))**. The crossover distortion is evident by inspecting the output voltage.

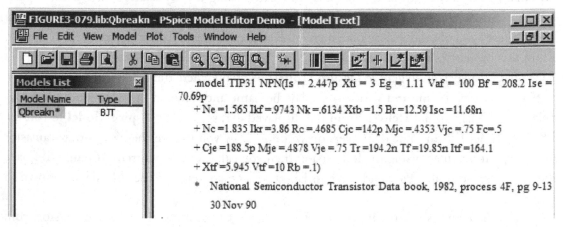

FIGURE3-079.lib:Qbreakn - PSpice Model Editor Demo - [Model Text]

File Edit View Model Plot Tools Window Help

Models List

Model Name	Type
Qbreakn*	BJT

.model TIP31 NPN(Is = 2.447p Xti = 3 Eg = 1.11 Vaf = 100 Bf = 208.2 Ise = 70.69p
+ Ne =1.565 Ikf =.9743 Nk =.6134 Xtb =1.5 Br =12.59 Isc =11.68n
+ Nc =1.835 Ikr =3.86 Rc =.4685 Cjc =142p Mjc =.4353 Vjc =.75 Fc=.5
+ Cje =188.5p Mje =.4878 Vje =.75 Tr =194.2n Tf =19.85n Itf =164.1
+ Xtf =5.945 Vtf =10 Rb =.1)
* National Semiconductor Transistor Data book, 1982, process 4F, pg 9-13
   30 Nov 90

FIGURE 8.12: Edit Instance Model (Text)

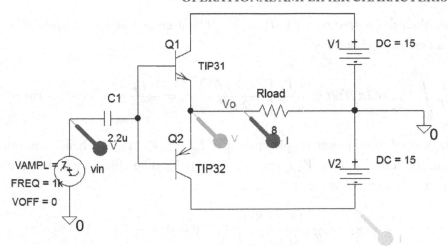

FIGURE 8.13: Push–pull amplifier

A power amplifier converts the DC input power to useful AC output power but in the process of doing this some power is dissipated in the circuit components. We need to consider the input power, the output power, and the power wasted in the circuit. The amplifier efficiency is calculated by considering the maximum input and output powers, and the maximum transistor power dissipation. A sinusoidal test signal is applied to the amplifier and we must consider the power in the load. The peak output voltage across the load R_L is V_p, then the output current is $I_p = V_p/R_L$. The maximum output power (RMS) is $P_o = \frac{V_p}{\sqrt{2}}\frac{I_p}{\sqrt{2}} = \frac{V_p^2}{2R_L}$. If we assume ideal

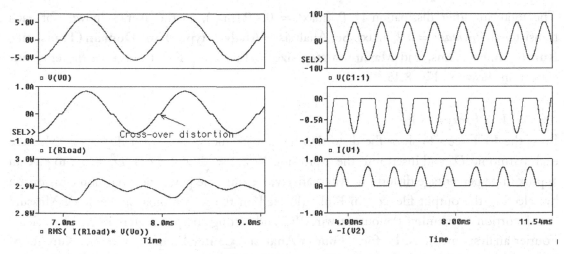

FIGURE 8.14: Class-B push–pull power amplifier

transistors, then the output power is $P_o = \frac{V_{cc}^2}{2R_L}$. The average value for a sinusoidal test signal over half a period $(T/2 = \pi)$ is

$$V_{\text{av}} = \frac{1}{T} \int\limits_0^{T/2} V_p \sin(2\pi ft)dt = -\frac{V_p}{T}\left[\frac{\cos(2\pi ft)}{2\pi f}\right]_0^{T/2} = -\frac{V_p}{2\pi}[\cos(\pi) - \cos(0)] = \frac{V_p}{\pi} \quad (8.6)$$

The average single-line power supply current is $I_{\text{av}} = (2V_p/\pi)/R_L$, hence the total average power from the source is $P_s = V_{cc}I_{\text{av}} = V_{cc}(2V_p/\pi)/R_L$. The efficiency is the ratio of output to input power as a percentage as

$$\eta = \frac{P_o}{P_s} = \frac{(V_p^2/2R_L)}{V_{cc}(2/\pi)(V_p/R_L)} \times 100\,\% = \frac{\pi}{4}\frac{V_p}{V_{cc}} \times 100\,\% \quad (8.7)$$

The maximum efficiency for class-B operation with $V_p = V_{cc}$ is $\eta = \frac{\pi}{4}\frac{V_{cc}}{V_{cc}} \times 100\,\% = 78.54\,\%$. The power dissipation for each transistor is the difference between the supply power and the power dissipated in the load:

$$P_D = P_s - P_o = V_{cc}\frac{2}{\pi}\frac{V_p}{R_L} - \frac{V_p^2}{2R_L} \quad (8.8)$$

Differentiate (8.8) with respect to V_p and equate to zero. This gives the maximum transistor power dissipation $P_D(\text{max})\frac{dP_D}{dV_p} = \frac{2}{\pi}\frac{V_{cc}}{R_L} - \frac{V_p}{R_L} = 0$. If $V_p = 2V_{cc}/\pi$, then (8.8) yields

$$P_D(\text{max}) = \frac{2}{\pi}\frac{2}{\pi}\frac{V_{cc}^2}{R_L}V_{cc} - \frac{4}{\pi^2}\frac{V_{cc}^2}{2R_L} = \frac{2}{\pi^2}\frac{V_{cc}^2}{R_L} \approx 0.2\frac{V_{cc}^2}{R_L} \quad (8.9)$$

The total amplifier dissipation is $P_D(\text{max}) = 0.4P_o(\text{max})$, so the average power for each transistor is $P_D(max) = 0.2_o(\text{max})$. Set Analysis to **Analysis type: Time Domain (Transient)**, **Run to time** = 3ms, and **Maximum step size** = 1us, Press **F11** to observe the crossover distortion shown in Fig. 8.15.

8.3.1   The Output File

Press the **FFT** icon to show the distortion harmonics due to cross-over distortion. Measure each harmonic D_n and insert into the expression %total = $\sqrt{D_1^2 + D_2^2 + D_3^2 + \cdots}$, to give an approximate percentage distortion. Alternatively, measure the percentage harmonic distortion by selecting the output file icon in Probe (located on the icon toolbar at the left). Measure the distortion by naming the output wire VO and setting the parameters in the Transient/Fourier analysis menu as: **Perform Fourier Analysis, Center Frequency** = 1k, **Number of Harmonics** = 10, and **Output Variables** = v(VO). The results of this analysis are found at

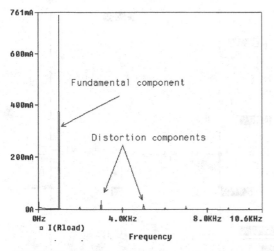
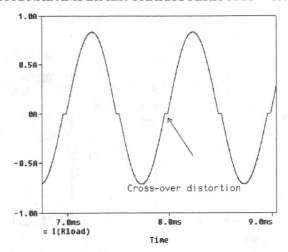

FIGURE 8.15: Crossover distortion

the end of the output file shown below (opened from the **Analysis** menu, or the **V̲iew** menu in Probe).

FOURIER COMPONENTS OF TRANSIENT RESPONSE V(VO)
DC COMPONENT = 7.156562E+00
HARMONIC FREQUENCY FOURIER NORMALIZED PHASE NORMALIZED
NO (HZ) COMPONENT COMPONENT (DEG) PHASE (DEG)

1	1.000E+03	4.970E+0	1.000E+0	4.770E−01	0.000E+00
2	2.000E+03	1.347E−03	2.711E−04	−1.506E+02	−1.511E+02
3	3.000E+03	8.964E−05	1.804E−05	−1.398E+02	−1.403E+02
4	4.000E+03	1.990E−04	4.004E−05	1.736E+02	1.732E+02
5	5.000E+03	1.631E−04	3.282E−05	1.764E+02	1.760E+02
6	6.000E+03	1.331E−04	2.677E−05	1.756E+02	1.752E+02
7	7.000E+03	1.139E−04	2.292E−05	1.748E+02	1.744E+02
8	8.000E+03	9.921E−05	1.996E−05	1.740E+02	1.735E+02
9	9.000E+03	8.806E−05	1.772E−05	1.737E+02	1.732E+02
10	1.000E+04	7.833E−05	1.576E−05	1.730E+02	1.725E+02

TOTAL HARMONIC DISTORTION = 2.805018E-02 PERCENT

Compare the *total harmonic distortion* of 2.8 % to that calculated using % distortion = $\sqrt{D_1^2 + D_2^2 + D_3^2 + ...}$. The previous circuit is modified to include a trickle bias as shown in Fig. 8.16. This quasi-complimentary push–pull amplifier has the output transistors connected in a Darlington pair configuration to reduce crossover distortion and yield lower output impedance values.

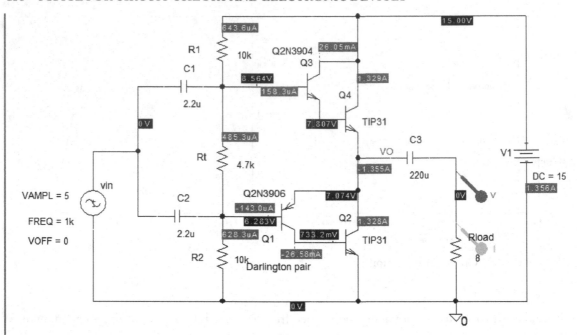

FIGURE 8.16: Improved power amplifier design

Make Rt variable in order to investigate crossover distortion. Change it from 47 kΩ to {**rvar**}, and place a **Param** part to define the name and value. The average power is plotted in Fig. 8.17 using the **AVG( )** operator from the **Trace/Add Trace** (or the insert button) function in Probe.

8.4   MOSFET DEVICE CHARACTERISTIC
Draw the MOSFET schematic shown in Fig. 8.18.

Set vdd to 5 V, and sweep vg from 0 V to 5 V to plot the transfer characteristic shown in Fig. 8.19.

Fig. 8.20 shows the output characteristic obtained by setting the **Primary Sweep** vdd from 0 to 5 V in steps of 0.1 V, and the **Secondary Sweep** vg from 0 to 6 V in steps of 0.5 V.

Draw the complimentary symmetry metal oxide (CMOS) schematic in Fig. 8.21.

The breakout symbol models **Mbreakp3** and **MbreakN3** are then associated with the two models listed in the comments in Section 8.4.1. Sweep vin from 0 to 5 V, in steps of 0.001, Press **F11** to simulate and plot the vin/vout characteristic shown in Fig. 8.22.

This characteristic shows that when the input level has a high positive value it causes the PMOS to turn off and the NMOS to turn on, thus making the output go to 0 V. When the

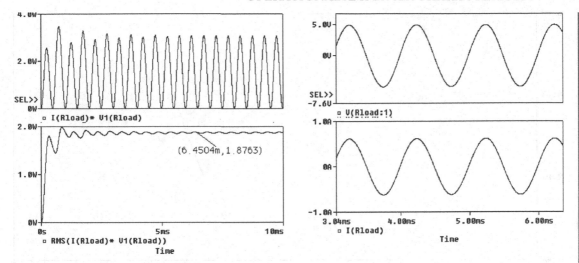

FIGURE 8.17: Average power output

input level is low, it causes the PMOS to turn on and the NMOS to turn off, which results in the output going high i.e. 5 V.

8.4.1   CMOS Model
The model for the MbreakP3 part is

```
.MODEL NCH NMOS (level=2 LD=0.15U TOX=200.0E−10 NSUB=5.37E+15
    + VTO=0.74 KP=8.0E−05 GAMMA=0.54 PHI=0.6 U0=656 UEXP=0.157 UCRIT=31444
    + DELTA=2.34 VMAX=55261 Xj=0.2U LAMBDA=0.037 NFS=1E+12 NEFF=1.001 NSS=1E+11
    + TPG=1.0 RSH=70.00
    + CGDO=4.3E−10 CGSO=4.3E−10 Cj=0.0003 Mj=0.66
    + CJSW=8.0E−10 MJSW=0.24 PB=0.58
```

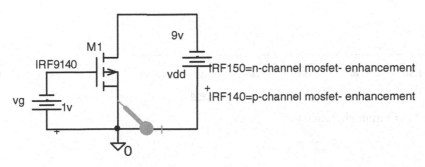

FIGURE 8.18: MOSFET circuit

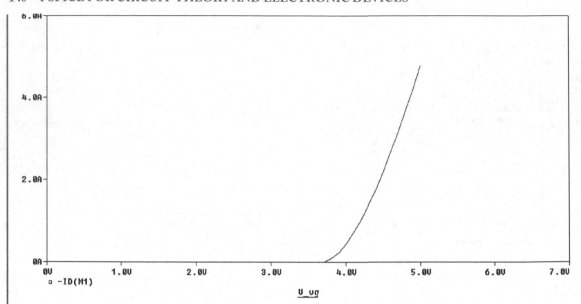

FIGURE 8.19: Transfer characteristic

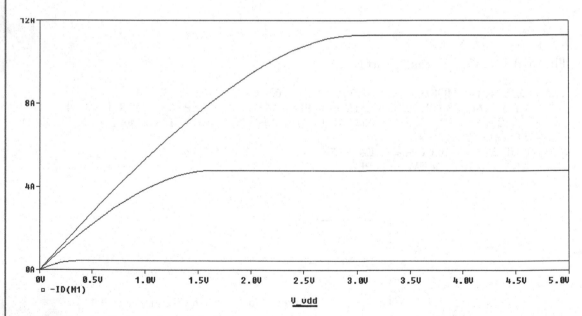

FIGURE 8.20: Output characteristic

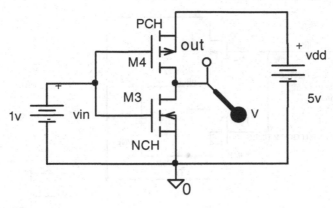

FIGURE 8.21: CMOS inverter

The model for the MbreakN3 part is

.MODEL **PCH** PMOS(level=2 LD=0.15U TOX=200.0E−10 NSUB=4.33E+15
  + VTO=−0.74 KP=2.70E−05 GAMMA=0.58 PHI=0.6 U0=262 UEXP=0.324 UCRIT=65720
  + DELTA=1.79 VMAX=25694 Xj=0.25U LAMBDA=0.061 NFS=1E+12 NEFF=1.001 NSS=1E+11
  + TPG=1.0 RSH=121.00
  + CGDO=4.3E−10 CGSO=4.3E−10 Cj=0.0005 Mj=0.51
  + CJSW=1.35E−10 MJSW=0.24 PB=0.64

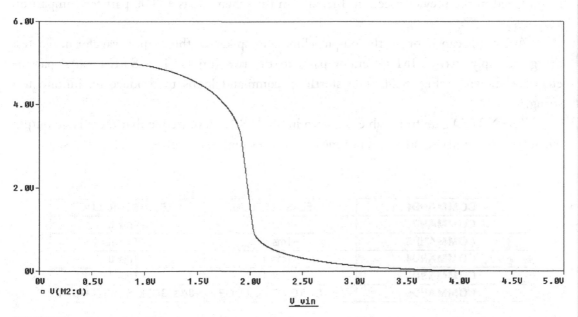

FIGURE 8.22: A plot of the input voltage versus the output voltage

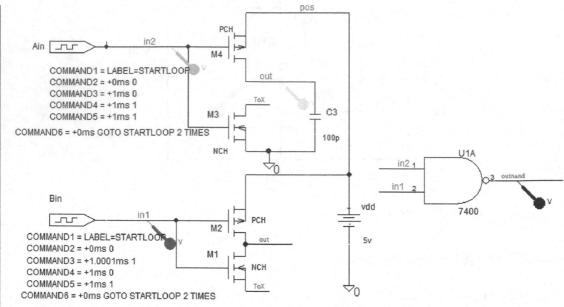

FIGURE 8.23: Two-input NAND gate

8.4.2   Nand Gate

Draw the two-input NAND gate schematic shown in Fig. 8.23 using the CMOS devices investigated in the previous section. Included in the schematic is a 7400 part for comparison purposes.

A 100 pF capacitor on the output eliminates spikes in the output waveform. To test the gate, apply two STIM generator parts to the two inputs. The **STIM** parts' parameters are shown in Fig. 8.24. The **startloop** command loops to produce an infinite test string.

The NAND gate truth table is shown in Table 8.1. Here we see that there is no output when the two inputs are identical but the output goes high for all other combinations.

COMMAND1	LABEL=STARTLOOP	LABEL=STARTLOOP
COMMAND2	+0ms 0	+0ms 0
COMMAND3	+1ms 0	+1.0001ms 1
COMMAND4	+1ms 1	+1ms 0
COMMAND5	+1ms 1	+1ms 1
COMMAND6	+0ms GOTO STARTLOOP 2	+0ms GOTO STARTLOOP 2

FIGURE 8.24: STIM1 and STIM2 parameters

TABLE 8.1: NAND Gate		
A	B	C
0	0	1
0	1	1
1	0	1
1	1	0

Comparing our model to an existing two-input 7400 NAND part shows how the "home-made gate" gives comparable results, as shown in Fig. 8.25.

8.4.3 Nor Gate

Draw the NOR gate schematic displayed in Fig. 8.26, and compare its performance to a standard 7402 NOR part.

The two-input NOR gate truth table in Table 8.2 shows an output present only when the two inputs are different.

Press **F11** to simulate and produce the gate waveforms that are displayed in Fig. 8.27.

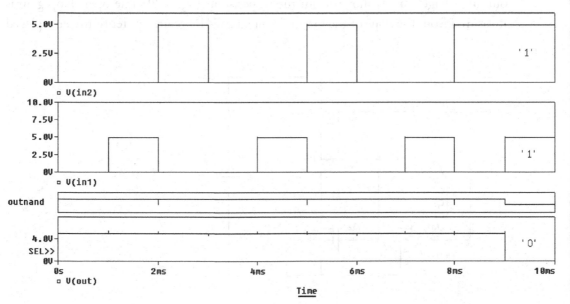

FIGURE 8.25: NAND gate waveforms

TABLE 8.2: Nor Gate		
A	B	C
A	B	C
0	0	1
0	1	0
1	0	0
1	1	0

8.5 EXERCISES

(1) Fig. 8.28 shows a schematic for obtaining the ua741 offset characteristics: The input offset voltage is the input voltage required to bring the output voltage to zero with zero signal voltage applied. The maximum, or short-circuit current output is obtained using a small output resistance (1 uΩ), and plotting the output current.

To measure the offset voltage (mV), make all the parameters of the **VSIN** part zero and do a DC sweep from −1mV to +1mV, in steps of 0.1uV. This should produce a graph showing the output voltage saturating at ± rail voltage, but with a positive sloping part in the middle of the plot. The offset voltage is the input voltage necessary to make the output voltage zero as shown from the response in Fig. 8.29. The open-loop gain is measured from the slope of the graph. In a real ua741 circuit, a potentiometer is placed

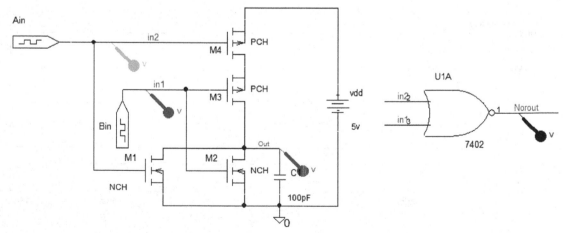

FIGURE 8.26: A two-input NOR gate

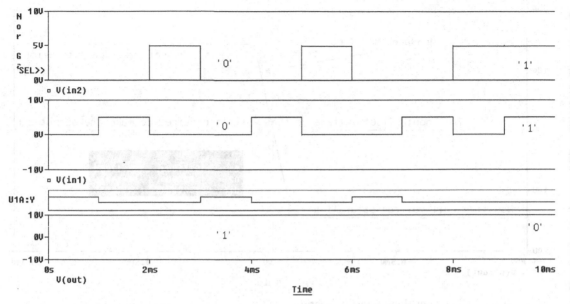

FIGURE 8.27: Two-input NOR gate waveforms

across the offset adjusting pins and then connected to the negative supply where it is adjusted to reduce offsets. These pins are displayed in the PSpice ua741 symbol, but are not actually modeled.

Measure the input bias current by pressing the "I" icon but with both inputs grounded to earth.

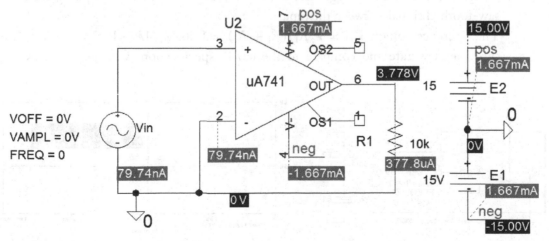

FIGURE 8.28: Open-loop measurements

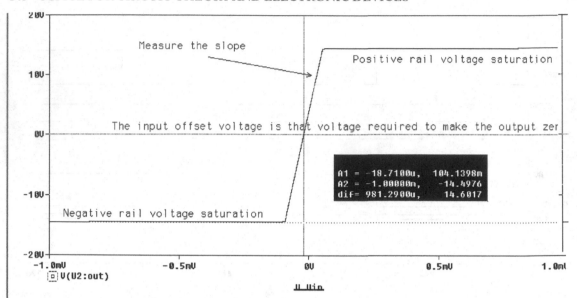

FIGURE 8.29: The open-loop transfer function

(2) To measure the input and output impedances, follow the procedure outlined in [ref: 3]. Measure the output impedance by injecting a voltage as shown and plotting the ratio of voltage/current as shown in Fig. 8.30.

(3) Investigate the opamp open-loop gain characteristics using the schematic shown in Fig. 8.31.

The response should be similar to that shown in Fig. 8.32.

(4) Investigate the Schmitt trigger circuit shown in Fig. 8.33. A **VPWL** part generates a saw-toothed signal to sweep the input.

The reference voltage $V_{\text{ref}} = V_o R_1/(R_1 + R_2) = 14.6.1k/(1k + 10k) = 1.3$ V. Measure the slew rate and compare to the ua741 specification. Change the x-axis to

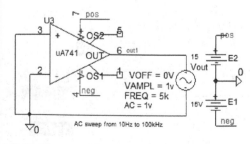

FIGURE 8.30: The output impedance

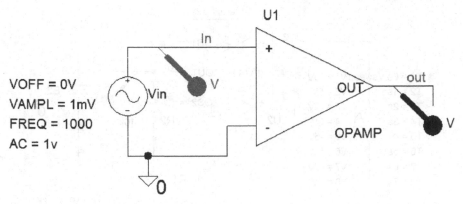

FIGURE 8.31: Noninverting op amp

Vin, and press **Alt PP** for a new upper plot axis in time as in Fig. 8.34 (select the **Plot/Unsynchronize X-Axis** menu for the upper plot). This is equivalent to changing the oscilloscope time base to the $X - Y$ mode and connecting a triangle wave from a function generator to the x-plates. Modify Fig. 8.33 to produce the circuit in Fig. 8.35. The hysterisis diagram shown in Fig. 8.36 is now plotted in the reverse direction when compared to Fig. 8.34. Add a noise generator in series with vsweep to verify the effectiveness of hysterisis.

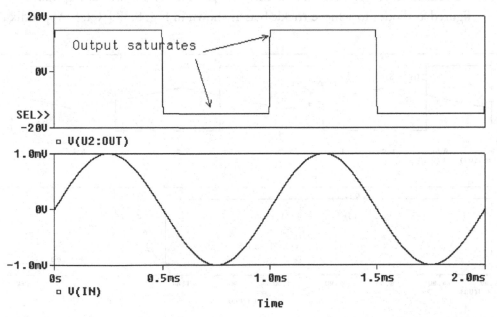

FIGURE 8.32: Gain-Bandwidth product

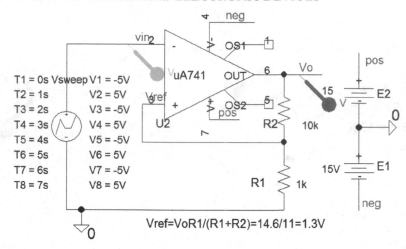

FIGURE 8.33: Schmitt-trigger circuit

(5) Investigate the astable oscillator using a ua741 IC in Fig. 8.37. The circuit to the right is for comparing the charging capacitors in each circuit.

The frequency of the square wave is calculated as

$$f_0 = \frac{1}{2CR_3 \ln[2(R_2/R_1) + 1]} \tag{8.10}$$

(6) To reduce crossover distortion, modify the previous circuit by adding transistors configured as diodes to achieve trickle bias as shown in Fig. 8.38 [Ref: 5 Appendix A].

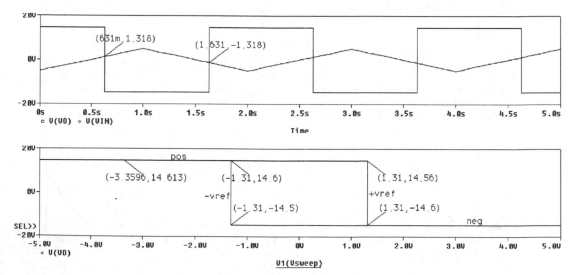

FIGURE 8.34: Schmitt trigger hysterisis

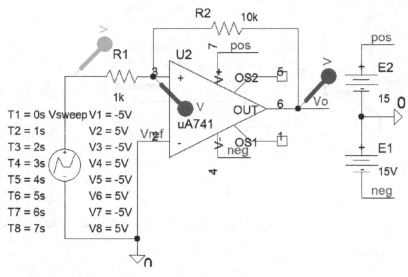

FIGURE 8.35: Schmitt trigger

The waveforms for this improved amplifier are shown in Fig. 8.39.

(7) Investigate the operational amplifier model using **ABM** parts shown in Fig. 8.40. The operational amplifier transfer function is entered in the L**aplace** part as a low-pass filter with a 30 rs^{-1} cut-off frequency (pole frequency $f_c = 4.78$ Hz). The **Diff** part sums the input signals and the **Gain part** represents the very large open-loop gain of 100 k.

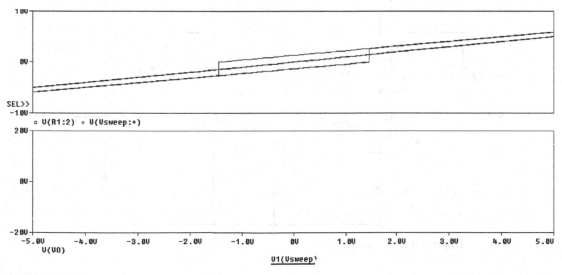

FIGURE 8.36:

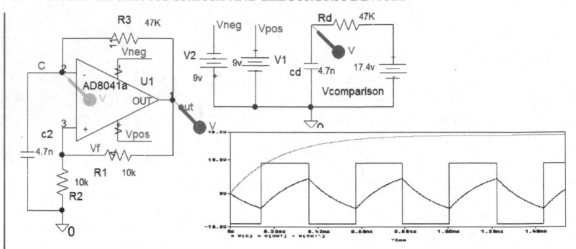

FIGURE 8.37: Astable oscillator

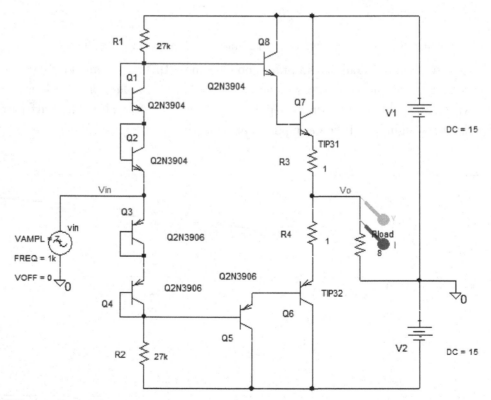

FIGURE 8.38: Improved power amplifier

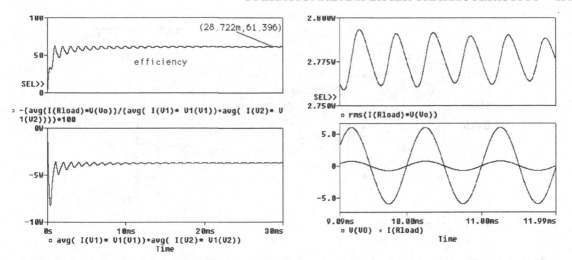

FIGURE 8.39: Power amplifier signals

The **limit** part limits the output to a value a little below the power supply value. We could use the **Glimit** part (contains the gain and limit parts) [ref: 4].

(8) The schematic in Fig. 8.41 uses the very good specification AD8041 opamp. Select the resistor R_2 and enter the value {**Rgain**}. Place and select a **Param** part from the **Special** library. Highlight **NAME1**, and enter Rgain in the **Value** box. Similarly, select the **VALUE1 attribute**, and enter 1k in the **Value** box. From the **Analysis Setup** menu, select**Parametric,** and in that sub menu, define **Name** = Rgain, **Start Value** = 1k,**End Value** = 100k, and **Increment** = 20k.

The frequency response in Fig. 8.41 shows how the AD8041 is a much superior op amplifier with a much larger passband frequency range compared to the very poor ua741 op amp frequency response.

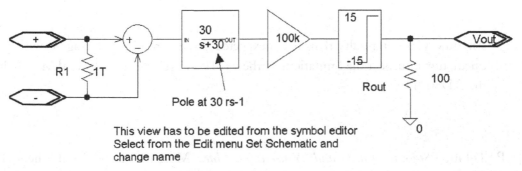

FIGURE 8.40: Constructing an opamp from ABM parts

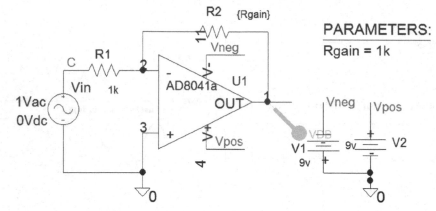

FIGURE 8.41: Inverting op amp

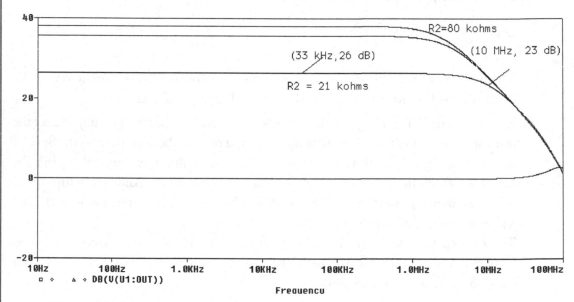

FIGURE 8.42: Frequency response

Anytime you get results that are unexpected, or the results don't agree with your calculations, suspect the limitations of the ua741 and substitute a better device such as the AD8041.

REFERENCES

[1]   P. Tobin, *PSpice for Filters and Transmission Lines*. Morgan Claypool publishers (Feb 2007).

[2] P. Tobin, (2007) *PSpice for Digital Signal Processing*. San Rafael, CA: Morgan Claypool, 2007.

[3] P. Tobin, *PSpice for Analog Communications Engineering*. San Rafael, CA: Morgan Claypool, 2007.

[4] M. E. Herniter Merrill, *Schematic Capture with PSpice*. New York: Macmillan, 1994.

[5] OrCAD PSpice for Windows, Volume II: Devices, Circuits, and Operational Amplifiers (3rd Edition) (Paperback) by Roy W. Goody New Jersey.

[6] Tobin Paul, *PSpice for Digital Communications Morgan Claypool publishers* Feb 2007.

APPENDIX A: LAPLACE AND z-TRANSFORM TABLE

FUNCTION	$f(t)$	LAPLACE TRANSFORM	$f(n)$	z-TRANSFORM ($t = nT = n$)
Unit step	$u(t)$	$\dfrac{1}{s}$	$u(n)$	$\dfrac{z}{z-1}$
Unit impulse	$\delta(t)$	1	$\delta(n)$	1
Unit ramp	T	$\dfrac{1}{s^2}$	N	$\dfrac{nz}{(z-1)^2}$
Polynomial	t^n	$\dfrac{n!}{s^{n+1}}$	t^n	$\dfrac{T^2 z(z+1)}{(z-1)^2}$ for n $= 2$
Decaying exponential	e^{-at}	$\dfrac{1}{(s+a)}$	$e^{-an}u(n)$	$\dfrac{z}{z - e^{-an}}$
Growing exponential	$\dfrac{1}{a(1 - e^{-at})}$	$\dfrac{1}{(s+a)(s)}$	$\dfrac{1}{a(1 - e^{-an})}$	$\dfrac{z(1 - e^{-an})}{a(z-1)(z - e^{-an})}$
Sine	$\sin(\omega t)$	$\dfrac{\omega}{(s^2 + \omega^2)}$	$\sin(n\theta)u(n)$	$\dfrac{z \sin n\theta}{z^2 - 2z \sin n\theta + 1}$
Cosine	$\cos(\omega t)$	$\dfrac{s}{(s^2 + \omega^2)}$	$\cos(n\theta)u(n)$	$\dfrac{z(z - \cos n\theta)}{z^2 - 2z \cos n\theta + 1}$
Damped sine	$e^{-at}\sin(\omega t)$	$\dfrac{\omega}{[(s+a)^2 + \omega^2)]}$	$e^{-an}\sin(n\theta)$	$\dfrac{ze^{-an}\sin(n\theta)}{z^2 - 2ze^{-an}\cos n\theta + e^{-2an}}$
Damped cosine	$e^{-at}\cos(\omega t)$	$\dfrac{(s+a)}{[(s+a)^2 + \omega^2)]}$	$e^{-an}\cos(n\theta)$	$\dfrac{z^2 - ze^{-an}\cos(n\theta)}{z^2 - 2ze^{-an}\cos n\theta + e^{-2an}}$
Delay	$f(t-k)$	e^{-sk}	$f(n-k)$	z^{-k}

Index

Author Biography

Paul Tobin graduated from Kevin Street College of Technology (now the Dublin Institute of Technology) with honours in electronic engineering and went to work for the Irish National Telecommunications company. Here, he was involved in redesigning the analogue junction network replacing cables with PCM systems over optical fibres. He gave a paper on the design of this new digital junction network to the Institute of Engineers of Ireland in 1982 and was awarded a Smith testimonial for one of the best papers that year. Having taught part-time courses in telecommunications systems in Kevin Street, he was invited to apply for a full-time lecture post. He accepted and started lecturing full time in 1983. Over the last twenty years he has given courses in telecommunications, digital signal processing and circuit theory.

He graduated with honours in 1998 having completed a taught MSc in various DSP topics and a project using the Wavelet Transform and neural networks to classify EEG (brain waves) associated with different mental tasks. He has been a 'guest professor' in the Institut Universitaire de Technologie (IUT), Bethune, France for the past four years giving courses in PSpice simulation topics. He wrote an unpublished book on PSpice but was persuaded by Joel Claypool (of Morgan and Claypool Publishers) at an engineering conference in Puerto Rico (July 2006), to break it into five PSpice books. One of the books introduces a novel way of teaching DSP using PSpice. There are over 500 worked examples in the five books covering a range of topics with sufficient theory and simulation results from basic circuit theory right up to advanced communication principles. Most of these worked example circuit have been thoroughly 'student tested' by Irish and International students and should mean little or no errors but alas. . . He married Marie and had four grown sons and his hobbies include playing modern jazz on double bass and piano but grew up playing G-banjo and guitar. His other hobby is flying and obtained a private pilots license (PPL) in the early 80's.

Printed in the United States
by Baker & Taylor Publisher Services